# Sociological Studies of
# Environmental Conflict

# Sociological Studies of Environmental Conflict

Sebahattin Ziyanak,
Mehmet Soyer, and Dian Jordan

**Hamilton Books**

Lanham • Boulder • New York • Toronto • London

Published by Hamilton Books
An imprint of The Rowman & Littlefield Publishing Group, Inc.
4501 Forbes Boulevard, Suite 200, Lanham, Maryland 20706
Hamilton Books Acquisitions Department (301) 459-3366

6 Tinworth Street, London SE11 5AL

British Library Cataloguing in Publication Information Available

**Library of Congress Control Number: 2019916098**

ISBN: 978-0-7618-7174-3 (pbk. : alk. paper)
ISBN: 978-0-7618-7175-0 (electronic)

∞™ The paper used in this publication meets the minimum requirements of American National Standard for Information Sciences Permanence of Paper for Printed Library Materials, ANSI/NISO Z39.48-1992.

# Contents

# Introduction

## *The Environmental Studies of Natural Resources*

## By Sebahattin Ziyanak,
## Dian Jordan, and Mehmet Soyer

The environmental studies about natural resource issues are often studied as conflicts; this book is carefully designed to expound on how resolutions are negotiated and maintained. A number of factors influence how conflicts are framed and how resolutions are determined regarding fracking, shared waters and environmental threats. This book explores the power, community activism, and politics regarding natural resources. Decisions often ignore ecological and social sustainability stewardship needs. By understanding how socio-political dynamics affect policy and negotiation, this book also contributes to the understanding of how natural resource policies are negotiated. It illuminates social inequalities between rural and urban populations.

We are living through images of the world around us that are generated by the media and that construct our understanding of politics and social problems (Gamson et al., 1992). Hence, local news media can become an important venue for grassroots groups to influence the public discourse about the possible impact of natural gas development, and water resource conflict. Since first-hand public discourse is likely to be minimal, through media, grassroots groups try to inform and educate people about environmental threats such as the greenhouse effect, climate change, ozone layer depletion, and water and air pollution to name a few. Such efforts have an influence on the general public's understanding and attitude toward their environment (Shanahan et al, 2015). We have witnessed that local newspapers play a more dynamic role in the civic life of local communities than many Americans believe. In the survey, local newspapers were mentioned as the most relied-upon source for information for all adults concerning local crime, taxes, government activities, schools, politics, job opportunities, community events, arts events, zoning information, social services, and real estate. Consequently, local newspapers continue to be key information sources to mobilize public opinion (Ali & Radcliffe, 2017, Rosenstiel et al., 2011).

The purpose of Chapter one through chapter four is to explain the debates surrounding environmental and public health concerns and the economic benefits of fracking. According to Boudet et al. (2016) those who support fracking live in areas where employment growth has grown due to the mining and natural resource sectors, are conservative politically and see the positive in the energy and economic supply outcomes. This book contributes to a better understanding of the role of local newspapers in the power-struggle between grassroots groups. This research studies local media's coverage and the discourse that claims-making groups generate to influence public opinion about an emerging social problem of gas drilling in the region. The integrated model of the theoretical framework to analyze the power struggle between two contending groups contributes to the social movement literature. The research addressed political debates weighing economic advantage against public health, for which advocates, and citizens alike turn to local news media and the Denton City Hall.

Chapter one aims to address two significant following questions: 1) What is the role of civil disobedience in New York activists' fight against fracking? 2) Why did Steingraber voluntarily accept a fifteen-day sentence for trespassing, rather than pay a fine? Chapter one discusses that the purpose of these letters which Steingraber uses writing to send a message beyond the jail is not to offer a clear path forward, but rather to grant credibility to the path she and others have already taken.

Throughout Chapter two and Chapter five, authors reviewed significant findings in relation to three overarching research questions. The first question was "How did campaign advocates from "Frack Free Denton" and "Denton Taxpayers For A Strong Economy" construct fracking in general?" The major claims constructed by both groups were glimpsed from the participants' interview. The second question was "How did each of these groups challenge the claims-making activities and goals of their adversaries?" The finding showed that both groups participated in the claims-making activities to win the power struggle over fracking during the campaign. The last question was "How did the local newspaper, the Denton Record Chronicle, become the field of power-struggle of grassroots groups (Frack Free Denton and Denton Taxpayers For A Strong Economy) over fracking?" The study revealed that the local news media became an arena of power struggle for the two groups. The supporters of both groups wrote letters to the editor and guest columns in support of their respective positions. Moreover, the claims-making activities of both groups find expressions in the news coverage.

Since the Frack Free Denton organization had more activities, their claims making actions received more media coverage than the fewer activities of the Denton Taxpayers for A Strong Economy group.

Throughout Chapter six and Chapter eight, this book demonstrates how historical documents can be utilized to identify patterns of politics and power related to water disputes, and in turn, how politics and power affect negotiation of water conflicts.

Sociology of water provides a framework from which to study the social construction of knowledge that is applied to water discourse. Agencies and actors are working within a culture of values and practices that are responsible for determining what knowledge is applied and accepted for water discourse.

There is no question that water is important to society. The importance of sociological philosophy and the science of water predate Socrates. Thales of Miletus Turkey (624–546 BC) is considered one of the seven sages of the Ancient World. Socrates recognized him as the first person to proclaim that water is the primary element of all things. He considered it the original element of the universe. And second to water is life, Thales proclaimed all had a soul (McKirahan, 2010). Because we have a soul, we have society. Society and environment cannot exist exclusive of each other.

As of 2019, the United Nations has eight primary millennium development goals identified to improve global society. The crucial role of water and water management is evident for these goals. With seventy percent of global freshwater usage directed at agricultural purposes, eradicating extreme poverty and hunger is closely aligned to food production. Access to clean drinking water has been essentially tied to combating HIV/AIDS, malaria, and other diseases, as well as reducing child mortality and improving maternal health. Universal primary education and promoting gender equality and empowering women are also related to obtaining better access to water, since many time-consuming water collection duties are fulfilled by girls and women. Environmental sustainability and global partnerships for development are critically tied to the importance of creating ecologically sound practices and practical policies for interdependencies of shared water sources (UN 2013a).

Social equity and justice are parameters that intersect all frameworks for understanding water governance (Alex, 2011). Increased environmental knowledge, particularly as it pertains to the importance of the role of water for good stewardship, has led to a heightened awareness to water policies and the associated implications for selected populations as it relates to those policies. Nearly half (46 percent) of the earth is covered by transboundary river basins which affect 148 countries. Nearly five hundred international agreements attempt to develop cooperative management and minimize conflicts related to shared waters. However, these arrangements only address about 40 percent of the basins (United Nations, 2014). By default, international negotiations are a negotiation of cultures. This is particularly significant when considering "more than half of all accessible global freshwater run-off is currently withdrawn by human intervention" (Bandyopadhyay, 2009, p. 148).

Sociology of water is part exploration into the parameters of social equity and justice. International socio-political relations are examined in the case study related to transboundary conflict between the United States and Mexico along the Rio Grande River. Issues of water conflicts are increasingly becoming items on global agendas. World societies are increasingly becoming urban based. By studying the effects of water demands and water transport from rural to urban areas, this chapter contributes to understanding the effects of increased urbanization.

Additionally, this section of the book explores water conflict between states. The urban population center of Dallas Fort-Worth Texas (Tarrant County) laid claim to waters of the Red River, the river that serves as the border between Texas and Oklahoma. Chapter seven reveals how the dispute was finally settled in the Supreme Court of the United States.

Finally, the complicated practices of racial inequalities pertaining to American Indians, who have unique and sovereign rights, are discussed in chapter eight. The United States government, namely the Bureau of Indian Affairs, maintains particular duties and obligations for the tribes. The federally recognized tribes of the Chickasaw Nation and the Choctaw Nation of Oklahoma are considered sovereign entities—they are a country unto themselves. Yet, they are residents of the State of Oklahoma. This chapter addresses the macro and micro conditions of policy and power related to the various federal, sovereign, state, and city entities that were involved in the selling of the water of Sardis Lake.

## ORGANIZATION OF THE BOOK

Chapter one centers on Steingraber's rhetorical strategies and an analysis of Steingraber's letters. In this rhetorical analysis of Steingraber's writings, the author studies how Steingraber makes the situation that voluntary incarceration is a prized mode of environmental support. This chapter introduces the reader to the discussion on the dangers of hydraulic fracturing or "fracking," a controversial method of extracting oil and gas from underground shale formations.

Chapter two studies the power struggle of two rival groups (Frack Free Denton and Denton Taxpayers For A Strong Economy) over fracking in Denton Texas. Chapter two includes a more detailed account of the natural gas industry that has grown rapidly, and how North Texas has become a major shale gas-producing area. This chapter addresses "how did each of these groups challenge the claims-making activities and goals of their adversaries?" Authors conducted and analyzed data from ten in-depth interviews from each side to compare concerns about fracking. Authors developed the model of merging the theoretical frameworks of value-conflict and social construction of social problems by examining the stages of awareness, policy determination, and reform in the battle over fracking. Finally, this chapter finds that the new theoretical framework model is germane to many features of claims," "claims-makers," and "claims-making" activities.

In chapter three, the authors address the specific question of what role did the local newspaper play in the power struggle between the two grassroots organizations: Frack Free Denton and Denton Taxpayers For A Strong Economy.

Chapter four explains that undoubtedly, there are noteworthy gaps in previous studies and in the science concerning how drilling will affect persons and societies. Dependable foundations of energy are dynamic for acquiring our most basic essentials and preserving our way of life. Unbalanced international markets resulted in enhanced struggles to increase production of national energy resources. A number of research projects examine evolving environmental health issues associated to the hydraulic fracturing processes. This chapter focuses on the public health issues related to this process, which are of excessive concern for healthcare workers.

Chapter five covers theoretical and implication of the findings related to the previous chapters. Moreover, this chapter provides an insight into environmental and media sociology. In addition to that, Soyer–Ziyanak's stages of a social problem model is expounded.

Chapter six examines the conflicts and power exhibited in a case study of United States and Mexico transboundary riparian watersheds. Interrelationships of the macro and micro structural orientations connected to the shared water for these two North American countries is studied.

Chapter seven highlights a case study that exposes the historical context of American Indian water rights for two tribes in Oklahoma and eventually whether Oklahoma and the tribes have beneficial Red River water rights over another state, Texas. This chapter also examines key structures of microscopic and macroscopic events and activities that link subjective and objective examples of behaviors, actions, policy and regulation. Through this analysis, we can detect the power controls and encounters explicitly encountered in the Texas and Oklahoma dispute concerning water provisions from the Red River.

Chapter eight elucidates the federal government's duties and the evolving political strength of Choctaw and Chickasaw tribal power as it relates to land and water rights. Topics of economic rights to sell water and cultural valuation of water are addressed. Sociology has begun to recognize alternative ways of knowing from voices that have historically been silenced, such as those from American Indians. In this chapter, these many aspects of sociology of water are examined through identifying the historical context and contemporary issues presented within the case studies.

## BIBLIOGRAPHY

Alex, P. (2011). The ripple effect. New York, NY: Scribner.

Ali, C., & Radcliffe, D. (2017). Small-market newspapers in the digital age. Retrieved from https://www.cjr.org/tow_center_reports/local-small-market-newspapers-study.php.

Bandyopadhyay, J. (2009). *Water, ecosystems and society*. New Delhi IN: Sage Publications India.

Boudet, H., Bugden, D., Zanocco, C., & Maibach, E. (2016). The effect of industry activities on public support for fracking. *Environmental Politics, 25* (4), 593–612.

Gamson, W. A., Croteau, D., Hoynes, W., & Sasson, T. (1992). Media images and the social construction of reality. *Annual Review of Sociology, 18*373–393. doi:10.1146/annurev.so.18.080192.002105.

McKirahan, Richard D. 2010. *Philosophy before Socrates: An Introduction with Texts and Commentary*. Hackett Publishing: Indianapolis IN.

Rosenstiel, T., Mitchell, A., Purcell, K., & Rainie, L. (2011). How people learn about their local community. Retrieved from http://www.journalism.org/2011/09/26/local-news/.

Shanahan, J., McComas, K., & Deline, M. B. (2015). Environment on television, and their effects. The Routledge handbook of environment and communication.

United Nations. (2014, Jan 22). Statistics. United Nations Inter-agency Mechanism on all Freshwater Related Issues. Retrieved from http://www.unwater.org/statistics/statistics-detail/en/c/211204/ .

United Nations (UN) Water. (2013a, Apr 3). Water Cooperation. *United Nations*. Retrieved from http://www.unwater.org/water-cooperation-2013/water-cooperation/en/.

*Chapter One*

# Civil Disobedience in Anti-Fracking Activism

## *Sandra Steingraber's Ironic Constructions of Incarceration*

## By Mollie K. Murphy

Sandra Steingraber is a biologist and activist in the contemporary environmental movement. Perhaps best known for her trilogy of books written in the tradition of Rachel Carson, Steingraber's work from 2010 onward has focused almost exclusively on the dangers of hydraulic fracturing or "fracking," a highly controversial method of extracting oil and gas from underground shale formations. Although she calls for a nation-wide ban on fracking and a rapid move away from fossil fuel dependency, much of her activism focuses on local issues affecting her home state of New York. On March 18, 2013, Steingraber engaged in civil disobedience for the first time when she and ten others were charged with trespassing for blocking a truck from entering company gates. The truck belonged to Inergy Midstream (now Crestwood Equity Partners), a storage and transportation company that purchased the salt caverns beneath New York's Seneca Lake in 2008. Crestwood had planned to repurpose the chambers for storage of methane and liquefied petroleum gases, which would later be used for fracking (Esch, 2013; Porter, 2013). Upon her arrest, Steingraber voluntarily chose a two-week incarceration sentence over paying a fine. In a press conference statement following the arrest, she emphasizes that her actions were, for her and other community members, a last resort after "[taking] every legal avenue to raise the serious health, economic, and environmental concerns associated with the Inergy plant" (Steingraber, 2013e, n.p.). During her fifteen-day sentence, Steingraber authored a series of "Letters from Chemung County Jail."

7

The letters gesture toward Martin Luther King Jr.'s famous "Letter from Birmingham Jail" and Henry David Thoreau's "Civil Disobedience." Steingraber explains that while her own situation greatly differs from activists of the past (e.g., incarceration was, for her, a choice), civil disobedience is nonetheless beneficial to contemporary environmental activists. In this rhetorical analysis of Steingraber's writings, I examine how she makes the case that voluntary incarceration is a valuable mode of environmental advocacy. Through two ironic constructions of her perspective attained behind bars, Steingraber posits jail as affording epistemic privilege—knowledge gained through experiencing systemic oppression—to incarcerated civil disobedients (Stillion Southard, 2014). First, Steingraber describes jail as offering activists a heightened awareness of the stakes involved in the environmental struggle. Second, she depicts jail as a place wherein activists can both realize and activate their agency. In the most constraining conditions, awareness and action are not only possible, but also more potent and powerful. Mirroring rhetorical theorist Kenneth Burke's (1969) formula of irony as "what goes forth as A returns as non-A," Steingraber distorts a familiar narrative of jail as a restriction of freedom and agency, arguing that it instead offers unique benefits to environmental advocates in particular (p. 517). In contrast to Thoreau's notion of freedom as a "state of mind," Steingraber emphasizes the visceral feelings of her own incarceration: "I am very aware of my physical self, and sense that my biological life in jail is part of my message" (Steingraber, 2013b, n.p.). Steingraber's depiction of life in jail simultaneously emphasizes the physical and epistemological effects of the experience. Incarceration is not merely a state of mind, but also a bodily state, thus allowing it to (ironically) strengthen both perception and agency. This analysis illustrates the connection between epistemology, rhetoric, and environmental activism by showing how Steingraber constitutes incarceration as a mode of action "worthy of credence" (Aristotle, 2007, p. 38). Attention to Steingraber's construction of her positionality in jail shows how irony can function as a means to validate specific types of advocacy. In rhetorical scholarship, irony's most common form—romantic—is understood as creating significant constraints on a rhetor's ability to advocate action. As Galewski (2007) explains, romantic irony calls status quo narratives into question (transforming "A" to "non-A"), thus "profess[ing] to offer an alternative to and escape from existing realities" (p. 87). Yet by stopping at questioning and distorting, romantic irony does not offer a new way of understanding a situation (e.g., the experience of incarceration). Thus, "the unalleviated freedom that it ushers in can prove just as paralyzing and oppressive as the system it sought to reject" (Galewski, 2007, p. 87). Scholars have suggested various strategies for overcoming romantic irony's constraints. Whereas Terrill (2003) argues that inciting action necessitates "the collapse of irony," Galewski (2007) and Parker (2008) argue that different forms of irony—

dialectical and tragic—can function help to clarify the ambiguities of romantic irony. In any case, there is agreement that romantic irony cannot function to promote action unless accompanied by specific rhetorical forms. The ironies of Steingraber's letters are consistently romantic; she avoids explicitly advocating incarceration, and maintains that no form of action is more worthy than others. She explains her own civil disobedience and voluntary incarceration as "a highly personal decision" and "an individual act of conscience" (Steingraber, 2013d, n.p.). She notes that "there is more to fear from our inaction than from the consequences of our actions," yet never tells the reader what to do (Steingraber, 2013d, n.p.). Despite this lack of clarity, I contend that romantic ironic constructions of Steingraber's voluntary incarceration function to justify rather than incite political action, and thus clarify incarceration as beneficial—perhaps crucial—to environmental activism. While romantic irony is ill equipped to *promote* action, it may serve as a viable means to validate it. In what follows, I first review irony as an ambiguous rhetorical form riddled with constraints and, in doing so, elucidate Steingraber's use of irony in the "letters." Next, I explicate how her ironic constructions of incarceration posit the experience as offering privileged knowledge to activists, thus making it a laudable mode of advocacy. Finally, the conclusion summarizes the contributions of this analysis to rhetorical understandings of irony in environmental advocacy.

## THE AMBIGUITIES OF IRONY IN ADVOCACY

Alongside synecdoche, metonymy, and metaphor, rhetorical theorist Kenneth Burke (1969) selected irony a "master trope." He argues that these four tropes are "not . . . purely figurative" but play a "role in the discovery and description of 'the truth'" (Burke, 1969, p. 503). Irony is the "perspective of perspectives," and its configuration is notably distinct from the other tropes. Whereas metaphor shows "A" in terms of "B" and metonymy (reduction) and synecdoche (representation) break down "A" according to its parts, Burke (1969) describe the "overall formula" of irony as "what goes forth as A returns as non-A" (p. 517). To constitute a distorted version of "A," the ironic rhetor must present two ways of seeing at once by bringing contradictory images into the same "field of vision" (Terrill, 2003, p. 220). This differs from the other tropes, which all rely on sequence. As Terrill (2003) explains, "[t]he tenor and vehicle of metaphor cannot be presented simultaneously or at the same conceptual distance from the auditor," and representation likewise precludes the equal presence of part and whole (p. 228). In contrast, irony "asks that two or several things be presented before the auditor in the same place at the same time," so that they may together create a meaning that inexactly resembles each term, instead relying on both to constitute the third, composite meaning (Terrill, 2003, p. 223; see also Hutcheon, 1994). Irony is permeated with ambiguities, making it arguably the most complex of Burke's master tropes. Much of irony's rhetorical force is concealed; it takes place in the unsaid as much as the said, yet "needs both to happen" (Hutcheon, 1994, p. 12). Irony's "edge" depends heavily on a rhetor's ability to balance this tension between obscurity and clarity (Karstetter, 1964). Terrill (2003) argues that a collapse of this oscillation marks the end of irony; for this reason, irony is ill equipped to inspire political action. Terrill (2003) evinces this through an assessment of Frederick Douglass's famous oration, "What to a Slave is the Fourth of July?," delivered to a predominately white audience in 1852 in Rochester, New York, just two years after the passing of the Fugitive Slave Act. Douglass drew on irony to showcase inconsistencies between the birth of the nation and the institution of slavery. Irony worked "first as a strategy through which to allow [Douglass's] white audience to recover the attitudes of the founders," but then shifted "to force his audience to acknowledge its inconsistencies" by offering them a rhetorical "tour" of the internal slave trade (Terrill, 2003, p. 225, 229). As he presents listeners with various scenes, Douglass avoids commenting. As Terrill (2003) notes, "If the ironist seems actively to be pointing out a temporal or hierarchical relationship between the two views, then the perspective is no longer ironic" (p. 220). Necessarily, then, Douglass's call for action coincides with a departure from irony. Whereas Douglass begins the speech by positioning his audience to "stand mute before the horrific inconsistencies of U.S. slavery," he concludes by spelling such silence as complicity (Terrill, 2003, p. 217). Thus, although irony can promote insight, Terrill argues that it prevents the rhetor from making clear moral judgments or advocating a course of action. Whereas Terrill argues that releasing the ironic form is necessary in calls to action, other scholars have suggested different ways to navigate its constraints. In her analysis of the "Black Manifesto," Parker (2008) argues that James Forman's strategic reversal of the tragic U.S. plot of race relations "provid[ed] a functional antidote to irony's indirection" (p. 338). Specifically, the "Black Manifesto" contained three "ironic registers"—different ironic forms working within the same text—that together laid out roles for blacks and whites in what Forman posited was the inevitable reversal of race relations (Parker, 2008). Whereas blacks would become empowered leaders, whites would take on a supportive role. Parker argues that the tragic nature of the manifesto's plot "combined predictable ironic elements to accomplish a dramatic reversal that provoked reflection and elicited an emotional response" (Parker, 2008, p. 327). In the "Black Manifesto," tragedy and the ironic, strategic reversal of roles combined to promote action. In special cases, irony can overcome its own constraints. Galewski (2007) evinces this potential in her analysis of Judith Sargent Murray's 1790 pitch for women's intellectual capacity. Galewski (2007) argues that dialectic or "humble" irony enabled Murray to overcome the pitfalls of romantic irony. Romantic irony is often conflated with irony in general; indeed, it is romantic irony that follows Burke's "A" to "non-A" formula. By transforming "A" to "non-A" but not to something else entirely (e.g., "B") romantic irony "does not offer a new meaning that can substitute in the place of the old," and thus leaves the audience with a distorted but unclear version of "A" (Galweski, 2007, p. 87). Echoing Terrill's (2003) concerns, Galewski (2007) explains that "this modality of irony professes to offer an alternative to and escape from existing realities, but the unalleviated freedom that it ushers in can prove just as paralyzing and oppressive as the system it sought to reject" (p. 87). For Murray, two romantic ironies encouraged her readers to reconsider women's potential to reason: she presented feminine features associated with stupidity (e.g., tendencies toward gossip) as ironic evidence of intelligence, and argued that women's education ironically wasted their feminine potential (e.g., by hindering their ability to secure a marital partner) (Galewski, 2007). Although these ironies disrupted assumptions about women, they left readers of *On the Equality of the Sexes* with no clear idea of how to improve women's situations. Galewski (2007) explains how a third, dialectical irony worked to overcome the limitations of these romantic forms. Comic in its approach, Burke (1969) explains dialectical irony as "based upon a sense of fundamental kinship with the enemy" (p. 514). Rather than rejecting status quo narratives, dialectical irony maintains faith in the existing grammar. Instead of merely distorting "A," dialectical irony creates a clear path forward by both reconfiguring *and* reiterating the common, original term. Indebted to "A," "non-A" becomes a new, clarified version of the transformed yet "A" (Galewski, 2007). Rather than "carv[ing] out the negative space of a silhouette like romantic irony, [dialectical irony] emerges out of this abyss of infinite possibility in order to draw a new profile for the world to see" (Galewski, 2007, p. 99). Galewski (2007) shows that Murray exemplified dialectical irony in her argument that a classical, masculinist education would secure women's ability to achieve their intellectual potential. This gave readers a specific vision of what was needed to address social injustice.

Though permeated with ambiguity and limited in its potential to promote action, romantic irony is not without benefit. When unaccompanied by other rhetorical "antidotes," it may be said to demand even more skill of the rhetor by asking them to sustain a complicated dance with the audience. In its covert, indirectly argumentative form (romantic), irony "requires the speaker for a moment at least, to put on a mask" (Karstetter, 1964, p. 168). This is evidenced in Steingraber's letters from jail wherein she initially portrays herself as "criminal" before flipping the script and targeting Inergy/Crestwood Equity Partners as the criminal. This "strategic moment of reversal" is key to prompting the audience to question the initial term at hand (e.g., "criminality") (Parker, 2008; Spencer, 2016). As Spencer (2016) explains, "Irony involves bringing the audience along in what appears to be a conventional narrative and then trapping the audience at the moment of peripety" (p. 535). In doing so, it invites audiences to abandon the narrative of the status quo in favor of the rhetor's (unclear) narrative (Spencer, 2016). Though it does not offer a clear direction forward, romantic irony brings a doubled perspective into focus, thus prompting "discovery" of knowledge (Parker, 2008; Terrill, 2003). In effect, it asks the audience to partake the in the resolution of an incongruity, though *how* to partake in resolution is unstated.

Steingraber's "letters from jail" rely predominately on romantic irony. Rather than negotiating the constraints of romantic irony with another rhetorical form, I argue that the purpose of these letters is not to offer a clear path forward, but rather to grant credibility to the path she and others have already taken. In this regard, the letters serve as an appeal to credibility, or *ethos*. By articulating her positionality (materially, as incarcerated and spiritually, as an activist) as ironic, Steingraber gives credence to anti-fracking activists' civil disobedience. Whereas jail—as "A"—initially appears as a place that constrains agency, Steingraber posits it as an ironic location wherein individuals have access to knowledge that improves activism rather than prevents it. Thus, the transformation of incarceration from "A" (constraining) to "non-A" (ironically enabling in its constriction) hinges on her portrayal of incarceration as affording a unique way of knowing.

## IRONIC ACTIVISM AS ETHOS

Although the March 2013 protest against Inergy targeted a particular local issue—underground gas storage—Steingraber's first arrest marked a shift in her mode of address as an environmental advocate concerned with a broad array of issues. Despite years of researching, writing, and speaking about the well-known effects of toxic chemicals (including those used in fracking), Steingraber (2013d) expresses frustration that "very little has changed" (n.p.). In one of her letters from jail, she states: "Here is what I am now convinced of: the oil, gas, and coal industries—and all the hydrocarbon carcinogens they produce and release—will not be dismantled by good data alone" (Steingraber, 2013d, n.p.). Without directing her audience to take any specific action, she indicates incarceration as a valuable form of activism. As noted at the outset, she also explains her own experience in jail as deeply physical in contrast to Henry David Thoreau's perception of "the confinement of his physical self [as] inconsequential" (Steingraber, 2013b, n.p.). Separated from the outside world, Steingraber posits jail as heightening awareness of her need for the sights, sounds, and smells of nature. In her various "Letters from Chemung County Jail," incarceration comes to represent an ironic physical and epistemological location for activists. Those *fighting* corruption end up behind bars, yet the experience affords them two key benefits: they can more readily observe the stakes involved in the struggle, which in turn heightens commitment to the cause, and they can activate their agency to effect change.

## Heightening Awareness and Commitment

In three of her "letters," Steingraber describes separation from the natural world as offering heightened insight to what is at stake. The human/nature separation experienced in jail becomes a premonition, allowing her a glimpse into the possibility of complete environmental degradation. Burke (1969) argues that irony is implicit in statements about prediction or inevitability. This is evidenced in ironic predictions that "the developments that led to the rise will, by the further course of their development, 'inevitably' lead to the fall" (Burke, 1969, p. 517). Steingraber uses irony in this predictive sense; jail—devoid of nature—offers a vision of the future if fracking moves forward. Burke (1969) explains that "different casuistries appear" when "one tries to decide exactly what new characters, born of a given prior character, will be the 'inevitable' vessels of the prior character's deposition" (p. 517). In Steingraber's letters, fracking—the supposed "bridge" to renewable energy—becomes a dangerous trap that will further entrench society in fossil fuel dependency. To help her audience visualize the "directional substance" (Burke, 1969) of the pro-fracking path, then, Steingraber "inevitably" turns to irony. This rhetorical move is well evidenced in Steingraber's descriptions of jail as similar to a world ravaged by fracking. Life behind bars is bright, loud, and devoid of any beauty or local, healthy foods. Fluorescent lights stay on all night, and the noises of nearby birds are drowned out by "[t]he buzzing, banging, [and] clanking of jail and the growled announcements of guards on their two way radios—which also go on all night" (Steingraber, 2013c, n.p.). This perfectly mirrors her description of fracking risks in her "Earth Day Letter from Chemung County Jail to Environmental Leaders." Here, she explains the serious dangers posed by fracking including air and groundwater contamination, sinkholes, and cavern collapse, but is also careful to note lesser issues that threaten quality of life (Steingraber, 2013a). "Along with the 24-hour light pollution from the industrial lighting of the drill rigs and the 24-hour noise from the compressors, this facility will fill our scenic highways with fleets of diesel trucks and send train cars of hazardous, flammable cargo over our rickety rail trestles" (Steingraber, 2013a, n.p.). Thus, as the incarcerated activist knows, jail is in many ways akin to a future world ravaged by fracking operations. Throughout the letters, Steingraber illustrates that jail not only affords a glimpse into the "fractured" future, but also shows the activist that this disastrous future is preventable. Though jail initially appears conventionally as a forced separation from the outer world ("A"), it "returns" as an unsettling warning. Without action to halt fracking in specific and fossil fuel usage in general, the outer world will become akin to incarceration, thus rendering jail obsolete, at least in some regards. This pattern is perhaps most pointedly illustrated in the "Crappy Mom Manifesto: Letter to Fellow Mothers from the Chemung County Jail," wherein Steingraber (2014b) states:

> [E]nforced extended separation from the natural world serves as a potent reminder of everything we depend upon the world for us to do. Five days without clouds, sky, stars, leaves, birdsong, wind, sunlight, and fresh food has left me homesick to the point of grief. I now inhabit an ugly, diminished place devoid of life and beauty—and this is exactly the kind of harsh, ravaged world I do not want my children to inhabit. (n.p.)

Because the natural world does *not yet* mirror incarceration, Steingraber's excerpts serve as much as motivators as they do warnings. As the noises of jail compete with the birds outside, Steingraber (2013c) assures herself and her readers: "But the world, I knew, was out there somewhere" (n.p.). It is within the play between her notions of the natural world, incarceration, and their composite prediction—the possibility of ecological collapse—that Steingraber showcases the benefits of incarceration. Ironic depictions show that activists incarcerated for civil disobedience tap into a heightened awareness of what is at stake. Accordingly, if individuals want to consider and speak out about the risks of fracking, "a jail cell is not an inappropriate place to do so" (Steingraber, 2013b, n.p.). In addition to framing incarceration as a glimpse into a devastating possible future, Steingraber also frames it as (for the civil disobedient) an ironically positive, stress-free experience in contrast to life in the outside world. The "stakes" involved in the struggle against fracking are both the possibility of the not-yet-realized ecological collapse and the possibility of a continuation of the status quo. In her "Letter from Chemung County Jail to Fellow Cancer Survivors," Steingraber merges the experiences of incarceration and cancer treatment to create yet another composite view of jail time. After briefly overviewing the digitization of fingerprints and mug shots, she explains the experience as uncannily similar to cancer check-ups. "And then it hits you: how exactly like looking at one's own breasts on a mammogram! Only this time: you already know the length of your sentence; it's far shorter than having cancer, and it doesn't involve the possibility of death" (Steingraber, 2013d, n.p.). Unlike cancer treatment, jail is not plagued with uncertainties. By describing jail as, for her, far less stressful than cancer, Steingraber makes the current outside world ironically *akin* to incarceration. Though the letter targets cancer survivors specifically, its message is likely relatable to many. Even for those who are not cancer patients or survivors, most individuals have a personal connection to someone who has survived or died from the disease. The National Cancer Institute estimated 15.5 million cancer survivors living in the U.S. in 2016, and projects an increase in this number in the near future (Cancer Statistics). Of course, this number does not account for those who did not survive. Through this link to cancer, Steingraber shows her readers that they are already strangely familiar with the ordeals of incarceration; the difference is that cancer is far more difficult. Fingerprinting is less scary than viewing colonoscopy results, and twenty-four-hour lockup is less constraining than a forty-five minute MRI scan. While incarcerated, "you can reacquaint yourself with the bible, you can do pushups, you can think" (Steingraber, 2013d, n.p.). In contrast to cancer, jail is less mentally and physically disabling. By depicting jail as in some ways preferable to an experience that plagues society, Steingraber shows her readers that their freedoms are already constrained, even without a jail sentence. Steingraber thus takes a familiar understanding of incarceration and flips it on its head; the common narrative is that the stakes in going to jail are loss of freedoms, yet she illustrates that her readers are already living such a reality. By framing jail ironically as a lesser threat to freedom than cancer, Steingraber reduces uncertainty about the experience, making it appear as a feasible form of activism. Importantly, she frames this as something she discovered while incarcerated—not beforehand. As a first time offender, Steingraber (2013d) states: "What I didn't expect . . . was how well prepared I was for jail by my prior experience as a cancer patient. As far as I can see, if you've ever spent time in a hospital, tethered to a catheter tube, you have the skills you need to cope with incarceration" (n.p.). Going to jail to fight fracking—a practice that relies on the use of carcinogens, which contaminate air, water, and soil—is thus sensible and familiar to the cancer survivor, but perhaps to others as well. Through this positive, ironic depiction of jail, Steingraber fosters identification with her readers by negotiating her "location" behind bars with the position her readers already, in many ways, inhabit.

## Activating Agency

In her "letters," Steingraber also ironically frames jail as a place wherein activists can activate their agency. Though confining, she paints incarceration as offering the conditions to find determination, practice patience, foster creativity, and have one's activist message be heard. In her "Letter from Chemung County Jail, Part 2," she forwards a familiar narrative of jail as a place of constrained agency. In New York, all new inmates are given tuberculosis tests by requirement. Waiting for results takes about three days. Steingraber (2013c) explains the waiting period: "Isolation means you are locked inside your cell with no access to the phone . . . no access to books . . . and, of course, no access to wi fi, cell phones, email or the internet" (n.p.).

She writes the "letter" on the back of an inmate request form. As she deals with these constraints on agency, she observes her fellow inmates overcome the same barriers. A woman she calls "Stingray" breaks her tooth and, after an officer dismisses her pain, responds by yelling defiantly: "[S]he stood at the iron door and called for pain meds, over and over in a voice that I use for rally speeches. Full oration. Projecting to the rafters. . . . She got her pain meds" (Steingraber, 2013c, n.p.). Despite living behind bars, Stingray used her voice to get what she needed. As Steingraber (2013c) concludes the letter, she likens her efforts to Stingray's.

> So, here I am, ringing the alarm bell from my isolation cell on Earth Day. May my voice be as un-ignorable as Stingray's. . . . The fossil fuel party must come to an end. I am shouting at an iron door. Can you hear me now? (n.p.)

Of course, the difference is that whereas Stingray used her voice to influence those who could literally hear it, Steingraber uses writing to send a message beyond the jail. Nonetheless, what she learns from Stingray is that jail does not preclude having agency over one's conditions. Aligning with Spencer's (2016) notion of irony as "bring[ing] audience[s] into a story or a way of seeing the world that the rhetor unhinges, often in a jarring way," Steingraber dismantles notions of jail as a place where activists can accomplish little (p. 522). Ironically, it becomes a place where activists can learn the extent of their agency. In a number of her letters, Steingraber (2014a) ties incarceration to maternal agency, configuring it as a place where women ironically can be "good mothers" by fighting for their children's future. She explains that in jail, she "became more fearless" (n.p.). Incarceration allowed her to realize her agency over her children's future; rather than falling victim to what she terms "well informed futility syndrome" (inaction due to paralyzing despair) (Steingraber, 2011), Steingraber activates her potential. In the "Crappy Mom Manifesto," she explains: "[J]ail *teaches* you how to stand up and fight inside of desperate circumstances" (Steingraber, 2014b, n.p.). Incarcerated, she learns "a skill set we all need" as other mothers encourage her to fight, even behind bars. "As Ashley scolded me last night, while passing a sharpened pencil through the bars, 'You can't just sit there for the next 14 days. Start fighting'" (Steingraber, 2014c, n.p.). Although she cannot be with her children and thus engage in traditional maternal practices, jail time is an ironic example of "good mothering" in that it represents an effort to fight for the future even in the most constraining circumstances. In Steingraber's letters, the status quo narrative of "good mothering" undergoes an ironic transformation. Although she is a "crappy mom" in the conventional sense (e.g., she missed every one of her daughter's cross country races), Steingraber (2014b) describes herself as fulfilling the most fundamental maternal duty: "[I]n cell 3, I'm doing what's required so that my kids have a future. Above all else, my job as their mother is to provide them that" (n.p.). Rather than pressing on with day-to-day tasks, she "refuse[s] . . . to give up on life" (Steingraber, 2014b, n.p.). Through this irony, the implicit message is that, given the state of the environmental crisis, incarceration is *the* place for parents and non-parents alike to exercise agency over the future. In the above examples, Steingraber's ironic depictions of jail function to legitimize it as an appropriate mode of activism. It is not merely a forced separation from the outer world. Through irony, she portrays it as much more. Jail offers the incarcerated activist with a glimpse into the unsettling future, but also with hope. It constrains agency, but in doing so allows for a full realization of potential. While incarcerated for fifteen days, Steingraber learned her resourcefulness and her bravery. As an epistemological and material location, jail affords the incarcerated activist with privileged knowledge, and is thus impossible to dismiss as trivial, unnecessary, or counterproductive. Steingraber's letters function to validate action rather than incite it. Not once does she collapse the irony by telling the reader what action to take. She emphasizes that she does *not* advocate going to jail. In her letter to cancer survivors, she states, "I'm not calling you to unlawful behavior. Civil disobedience is a highly personal decision and, for me, came as an individual act of conscience—but I do contend that there is more to fear from our inaction than from the consequences of our actions" (Steingraber, 2013d, n.p.). She also emphasizes that incarceration is no more worthy of credence than other forms of advocacy, as "[a]ll roles are equally vital" (Steingraber, 2014a, n.p.). In balancing obscurity and clarity—jail is a praiseworthy form of activism, but not one she explicitly advocates—Steingraber sustains romantic irony, thus seemingly creating constraints for her own advocacy. Yet for Steingraber, the purpose of sustaining irony in the letters is not to advocate political action, but rather to bolster her credibility as someone behind bars. Her reconfigured construction of incarceration functions as an appeal to *ethos* for, as Skinner (2009) notes, it is difficult to condemn a rhetor's character when "in the course of the speech the audience begins to see themselves as sharing her motivations for engaging in behavior they might otherwise denounce as radical" (p. 242). Like many of her readers, Steingraber is concerned for her children's future, has battled cancer, and deeply values a healthy community and planet. Ironically, jail offers a location wherein one can more keenly realize the implications of toxic chemicals and fight against the fossil fuel industry.

## CONCLUSION

A year and a half after Steingraber's first arrest, Crestwood Equity Partners received federal approval to transform the salt caverns beneath Seneca Lake into storage containers for methane. One day before the company was authorized to begin construction, the "We Are Seneca Lake" (WASL) campaign held their first protest. Steingraber was among the many Seneca Lake "defenders," who cited drinking water contamination as a primary concern. Activists also vehemently opposed Crestwood's plans to become a gas storage hub for the entire northeast and thus enable fracking well beyond the confines of New York. Within two years of WASL's first protest, over 600 arrests had taken place (The Seneca Lake Defenders, 2016). What started as an eleven-person effort had blossomed into much more. In May 2017, Crestwood announced that it would be abandoning their plans to store methane in the caverns (Murray, 2017; Sonken, 2017). What is the role of civil disobedience in New York activists' fight against fracking? Why did Steingraber voluntarily accept a fifteen-day sentence for trespassing, rather than pay a fine? Why did she write letters from jail? Of course, activists have long drawn on civil disobedience to convey their dedication to a cause and the seriousness of an issue. Steingraber echoes this, arguing that filling jail cells with activists "provoke[s] a crisis that cannot be ignored by media or political leaders" (2014c, n.p.) and "shows seriousness of intent" (2014a, n.p.). However, I argue that media pull cannot fully explain Steingraber's voluntary incarceration. The majority of activists who were arrested did not serve jail time, yet the campaign still gained positive media coverage (Kaplan, 2014; McKinley, 2014).

Analysis of Steingraber's letters shows how she constitutes jail time as a legitimate and beneficial mode of advocacy. Irony is key to this appeal to both *ethos* (credibility) and *logos* (logic and good reason). Steingraber frames incarceration as an ironic experience wherein activists who are physically constrained can access beneficial knowledge. While behind bars, activists can fully realize the stakes involved in the anti-fracking struggle, and can access their potential to fight against a potentially devastating future. Steingraber thus illustrates the epistemological and material benefits of incarceration. Whereas previous analyses of romantic irony consider it as incompatible with political claims for action, I have reconsidered the assumed temporal relationship between irony and action. Romantic irony cannot offer audiences with a clear path forward, but when used as a means to legitimate rather than incite specific modes of advocacy, it offers activists with a powerful rhetorical resource. Though voluntary incarceration may seem frivolous, unnecessary, or even unethical, Steingraber draws on irony to distort this familiar narrative and produce a new, ironic understanding of time spent behind bars.

As noted at the outset, Steingraber's rhetorical strategies have changed dramatically over the past twenty years. Formerly a careful writer and poet, she is now an outspoken civil disobedient. Szasz (1994) observes that activism tends to be a radicalizing experience, especially for leaders. Although

Steingraber's contemporary rhetoric is certainly not radical in comparison to groups such as Earth First!, it is far less palatable than the writing of in her three books. Her advocacy has led her to believe that words alone will not suffice to change the status quo guiding the planet toward ecological disaster. Accordingly, Steingraber has narrowed her audience, placing her efforts toward galvanizing those who are already likeminded. One way in which she does so is by validating her own actions and, in doing so, offering a blueprint for others to act in ways that might at first seem unsavory or unnecessary, but offer significant benefits to activists.

## BIBLIOGRAPHY

Aristotle. (2007). *On rhetoric: A theory of civic discourse* (G. A. Kennedy, Trans.). New York, NY: Oxford University Press.

Burke, K. (1969). *A grammar of motives.* Berkeley, CA: University of California Press.

Cancer Statistics. (n.d.). *National Cancer Institute at the National Institute of Health.* Retrieved from https://www.cancer.gov/about-cancer/understanding/statistics

Esch, M. (2013, May 5). Gas-storage plans in NY's Finger Lakes draw outcry. *Yahoo News.* Retrieved from https://www.yahoo.com/news/gas-storage-plans-nys-finger-143559346.html

Galewski, E. (2007). The strange case for women's capacity to reason: Judith Sargent Murray's use of irony in "On the Equality of the Sexes" (1790). *Quarterly Journal of Speech, 93*(1), 84–108.

Hutcheon, L. (1994). *Irony's edge: The theory and politics of irony.* New York, NY: Routledge.

Kaplan, T. (2014, Dec 17). Citing health risks, Cuomo bans fracking in New York State. *The New York Times.* Retrieved from http://www.nytimes.com/2014/12/18/nyregion/cuomo-to-ban-fracking-in-new-york-state-citing-health-risks.html

Karstetter, A. B. (1964). Toward a theory of rhetorical irony. *Speech Monographs, 31*(2), 162–178.

McKinley, J. (2014, December 25). What pairs well with a Finger Lakes White? Not propane, vintners say. *The New York Times.* Retrieved from https://www.nytimes.com/2014/12/26/nyregion/new-york-winemakers-fight-gas-storage- plan-near-seneca-lake.html

Murray, J. (2017, May 10). Crestwood drops Seneca Lake natural gas storage plans. *The Star Gazette.* Retrieved from https://www.stargazette.com/story/news/2017/05/10/crestwood-drops-seneca-lake-natural-gas-storage-plans/316228001/

Parker, M. (2008). Ironic openings: The interpretive challenge of the "Black Manifesto." *Quarterly Journal of Speech, 94*(3), 320–342.

Porter, C. (2013, April 22). Environmentalist Sandra Steingraber in jail for fracking protest. *The Toronto Star.* Retrieved from https://www.thestar.com/news/world/2013/04/22/environmentalist_sandra_steingraber_in_jail_for_fracking_protest_porter.html , n.p.

Skinner, C. (2009). "She will have science": Ethos and audience in Mary Gove's "Lectures to Ladies." *Rhetoric Society Quarterly, 39*(3), 240–259.

Sonken, L. (2017, May 31). Is the Crestwood Midstream project truly dead? *Ithaca.com.* Retrieved from http://www.ithaca.com/news/ithaca/is-the-crestwood-midstream-project-truly-dead/article_6d903c98-4621-11e7-962a-f77d58b3eecb.html, n.p.

Spencer, L. G. (2016). Bishop Leontine Turpeau Current Kelly: Toward an ironic prophetic rhetoric. *Western Journal of Communication, 80*(5), 519–538.

Steingraber, S. (2013a, April 22). Earth Day letter from Chemung County Jail to environmental leaders. *Knowledge Town.* Retrieved from http://steingraber.com/earth-day-letter-from-chemung-county-jail-to-environmental-leaders/

Steingraber, S. (2013b, April 18). Letter from Chemung County Jail, part 1. *Knowledge Town.* Retrieved from http://steingraber.com/letter-from-chemung-county-jail-part-1/

Steingraber, S. (2013c, Apr 19). Letter from Chemung County Jail, part 2. *Knowledge Town.* Retrieved from http://steingraber.com/letter-from-chemung-county-jail-part-2/

Steingraber, S. (2013d, Apr 23). Letter from Chemung County Jail, part 3, a message to fellow cancer survivors. *Knowledge Town.* Retrieved from http://steingraber.com/letter-from-chemung-county-jail-part-3-a-message-to-fellow-cancer-survivors/

Steingraber, S. (2013e, Apr 17). Prepared statement for press conference. *Knowledge Town,* Retrieved from http://steingraber.com/sandra-steingraber-prepared-statement-for-press-conference/

Steingraber, S. (2011). *Raising Elijah: Protecting our children in an age of environmental crisis.* Boston, MA: De Capo Press.

Steingraber, S. (2014a, Nov 30). The case for going to jail and how to do it: Guide to the Chemung County Jail—for women. *Knowledge Town.* Retrieved from http://steingraber.com/the-case-for-going-to-jail-and-how-to-do-it-guides-to-the-schuyler-county-and-chemung-county-jails-for-seneca-lake-defenders/

Steingraber, S. (2014b, Nov 23). The crappy mom manifesto: Letter to fellow mothers from the Chemung County Jail. *Knowledge Town.* Retrieved from http://steingraber.com/the-crappy-mom-manifesto-letter-to-fellow-mothers-from-the-chemung-county-jail/

Steingraber, S. (2014c, Nov 21). Why I am in jail. *Knowledge Town.* Retrieved from http://steingraber.com/why-i-am-in-jail-2/

Stillion Southard, B. A. (2014). A rhetoric of epistemic privilege: Elizabeth Cady Stanton, Harriot Stanton Blatch, and the educated vote. *Advances in the History of Rhetoric, 17*(2), 157–178.

Szasz, A. (1994). *Ecopopulism: Toxic waste and the movement for environmental justice.* Minneapolis, MN: University of Minnesota Press.

Terrill, R. E. (2003). Irony, silence, and time: Frederick Douglass on the fifth of July. *Quarterly Journal of Speech, 89*(3), 216–234.

The Seneca Lake Defenders. (2016). *We Are Seneca Lake.* Retrieved from http://www.wearesenecalake.com/seneca-lake-defendes/.

## Chapter Two

# A Power Struggle over Fracking in Denton, Texas

## By Mehmet Soyer and Sebahattin Ziyanak

The Barnett Shale in North Texas is one of the largest natural gas fields in the United States. Since Denton County is one of the core counties with active natural gas drilling, the fracking industry has come under greater scrutiny by local activist groups, national environmental groups, oil and gas industry, and State lawmakers. Hydraulic fracturing (Fracking) is a method to extract natural gas in the rock formation. First, fracking wells are drilled vertically about hundreds to thousands of feet below. Then the drill continues horizontally about the feet. To be able to extract the natural gas, fracturing fluid, including water and other chemical additives, are pumped to open fractures in the rock formation. The injection fluid is stored in the tanks before recycling once the infection procedure is completed (EPA, 2018).

We have been lucky to find ourselves at ground-zero of the fracking debate in Denton. When we started this project, the subject matter was local. However, the subsequent power struggle attracted national, even global attention to this small town in north central Texas. The election on November 4th, 2014 made Denton the first city in Texas banning fracking. It became a common expression that the fracking controversy put the city of Denton on the map. The national and international press has also paid great attention to the issue.

The first and second author were both enrolled in a doctoral course in Denton. The first author also lived in Denton for two years. The fracking issue was exceedingly affecting media, Denton citizens, and social researchers. We believe that this is one of the most interesting subjects to engage. However, we would attempt to offer boarder and more impartial view of this significant issue since many things about what has been purported against fracking we have had with the negative view. We would like to simply point that we personally see this study as an opportunity to demonstrate all the questions and concerns from the both sides. Thus, unlike previous studies, the researchers' emphasis are on the campaign of the two grassroots groups on each side of the debate. In this study, our goals are to explore the way in which campaign advocates from each contending groups constructed fracking for the public. Emerging from the actual experiences of the researchers and local people in Denton, the power struggle of two grassroots groups will maintain their conflict from each other.

## FRACK FREE DENTON (FFD)

The Denton Drilling Awareness Group (Denton DAG) is a non-profit organization of citizens who are dedicated to informing the community about the potential threats of fracking on public health, the environment, and real estate values in the city of Denton. Moreover, this grassroots group introduced a residents' petition to ban hydraulic fracturing within the Denton city limits until the drilling has been proven harmless for environment and public health. The Denton Drilling Awareness Group (DAG) is incorporated as a non-profit educational group acknowledged by the State of Texas.

However, for FFD members, the feeling of accomplishment did not last long. The fracking ban was almost immediately overturned by the state legislature in Austin, Texas. The drilling companies resumed activity in the wells in Denton. Then the anti-fracking passion was reignited and the activists from the FFD group started their struggle once again. Some members even engaged in civil disobedience, blocking one of the well sites. We have seen that the stages of the natural history of social problems can be repeated. Denton residents are displeased with the state law and the reopening of drilling decision. In addition to that, due to health concerns, environmental groups such as FFD, advocates, and scientists disapproved hydraulic fracking. However, they started searching how to send a message to local citizens and how to educate them about the public health concerns pertaining to fracking to curb fracking operations. FFD's main emphasis is direct awareness on the potential negative effects of fracking. FFD arranged to express their voice around schools, public parks, and residential areas to get public support. The power struggle over fracking in Denton has the potential to trigger another phase of Awareness, Policy Determination, and Reform. As of 2019, the FFD activists have not given up. They came together to create the Texas Grassroots Network for the restoration of local control of natural resources.

## DENTON TAX PAYERS FOR STRONG ECONOMY (DTPSE)

Denton Tax Payers is pro-drilling interest group, collecting more than 8,000 signatures in Denton city supporting responsible local regulation instead of an "arbitrary and unconstitutional fracking ban." This interest group supports fracking in order to stimulate the economic development in Denton. They support property owner and mineral owner rights. Moreover, they assert that they protect taxpayers from future estate tax increases due to fracking ban.

## REVIEW OF THE LITERATURE

### Social Constructionism

The social construction of social problems is a complementary theoretical framework that explains how social phenomena are depicted as social problems through claims-making activities. Social problems are generated through claims-making activities (Best, 1995). The claims-makers construct the social problem (i.e., claim) through claims-making activities that aim at shaping the public perception.

Jones, Hillier, and Comfort (2013) examine about the public perception concerning fracking at United Kingdom. In this study, they argue that local opponent groups were well-mobilized and effectively utilized communication technologies and social media. Their findings demonstrate that the public relations and media outlets had a key role to play in achieving the contending groups (Jones, Hillier, & Comfort 2013).

Julios (2015) utilizes the notion of natural history of a social problem to examine the way in which honor killing has become the center of the UK government's policy agenda. In her study, the claims-making activities of the grassroots groups helped the government shape the policy about the honor violence. Linton (1991) applies the natural history of the social problems to examine the 1900s when young workers were considered as an official social problem in the Imperial Germany.

Best (1995) defines social problems as developing through the struggles of claims-makers who carry out concerns to public consideration. By characterizing a problem and portraying it as a specific category, claims-makers can form policymaking and civic reaction to the issue.

## THEORETICAL FRAMEWORK

The core of the awareness stage is problem-consciousness: the emergence of awareness of an unofficial and undefined social problem. The core of the awareness stage is problem-consciousness: the emergence of awareness of an unofficial and undefined social problem. The origin of every social problem dwells in the awakening of public in a given neighborhood to a recognition that assured cherished tenets are exposed by circumstances that have turned out to be severe (Fuller & Myers, 1941). The very first spark of awareness is hard to capture; however, as individuals come together for shared values and interests, awareness toward the condition arises. These members of society express their concerns in a measurable and observable form to inform others in the society. The message is that something should be done. However, there is no exact definition of the condition or a solution to the problem. Therefore, the individuals engage in unsynchronized random behaviors and are in a state of protest. The next move of these individuals or groups is to create common value or interest groups. These groups seek to raise consciousness about the potential threat to common values. Official complaints to press and civic authorities take place in order to gain attention to the social problem and move onto the next stage of social problem development (Fuller & Myers, 1941).

## Policy Determination

The policy determination stage starts when the debate over policy implications causes a conflict of interest. At this stage, opposing grassroots groups are constructed and each one takes action to reach out and create broader awareness from institutions such as health organizations, police departments, universities, and media.

This stage consists of three interrelated levels (Fuller & Myers, 1941). The first level is discussion by neighbors and other concerned individuals but in unorganized groups. The second level is discussion by interest groups and grassroots groups such as environmental groups, taxpayers, parent-teacher associations, women's clubs, and men's clubs. Finally, there is discussion among specialists and administrators in government or quasi-governmental units: the police departments, health officials, city council, social workers, and school boards. Therefore, these three interrelated stages characterize the dynamics of policy determination (Fuller & Myers, 1941).

## Reform

The last phase of the natural history of a social problem is reform. In the previous phase, the policy plan is developed and became the action plan. Now the action is under the administrators' responsibility. The action has two sides though; one is the public stage and the second one is the private stage. Since action is exercised to protect the shared values of one or more parties, general policies of the specific social problem have been discussed and described by interest groups and experts. In addition, there is still a probability of complicated legal issues to be cleared out before the action can be applied (Fuller & Myers, 1941).

In this phase, Fuller and Myers (1941) state that policy inquiries may be removed from the hands of the administrators whenever the general community cognizes its controls of censorship, rejection or vote. The emphasis is on the fact that this and that are being done. Institutionalization of the social problem makes this stage unique. Now that the policies are initiated by authorized policy enforcement agencies, the public agencies may prove to be sufficient or new agencies can be necessary in the face of another social problem. Fuller and Myers (1941) point out that, theoretically speaking, each stage marks itself off of its predecessor. However, it doesn't mean that the stages cannot coexist. The stages are not mutually exclusive; therefore, the development of the social problem can contain characteristics of each stage at any specific time.

In this theoretical framework, there are transitory stages in the natural history of a social problem. Transitory stages are not pre-determined. In other words, there are no guarantees that a social problem will move from the first to the second, or from the second to the third stage. It will continue if the parties can mobilize enough resources to push the process to the next stage. Therefore, the continuation of the process is contingent upon the longevity of the power of claims-makers.

## Fracking

Brasier et al. (2011) found out that wealth creation, job creation, increased business activity, and tax revenue are the four positive local economic im-

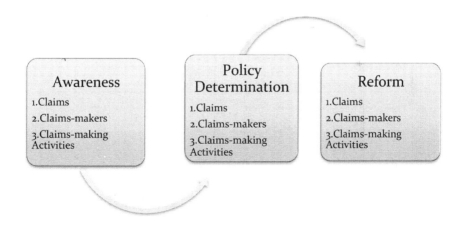

**Figure 2.1.**

pacts of natural gas drilling. Yet they also underline that residents experience community dissatisfaction and feelings of alienation (Brasier, et al., 2011). In the case of fracking in Denton, the issue is constructed as a community-level health threat. Gullion's (2015) study states the following:

> There is a flurry of discursive practice after the identification of the object of harm. The response from governmental officials is either minimal or confusing. Perceptions of the event vary, and risky is disputed. Grassroots activity responds to the threat in some manner. Talismans are used to help mitigate the risk. (p. 175)

Some sources focus on the impacts of environmental health problems from fracking. As reported by a Health Impact Assessment (HIA) in Colorado, there are eight environment-related health issues. These are air emissions, water contamination, truck traffic, noise and light pollution, accidents and malfunctions, strain on health care systems, psychosocial stress associated with community changes, and housing value depression (Witter et al., 2013). Moreover, some studies emphasized the possible air and water pollution impacts of gas drilling (Gullion, et al., 2011; Rabe & Borick, 2011). In this report, residents in Flower Mound, TX expressed their concerns that benzene contamination in the air as a result of fracking was causing cancer.

In the case of fracking in Denton, the first place to experience the controversy about fracking was Robson Ranch area, called an active adult luxury retirement community. Interestingly, gas companies built social facilities to attract people of the neighborhood. Most of the drilling was taking place in the western parts of the city of Denton that were still rural and sparsely populated. However, even here some residents from the neighborhood along with dozens of doctors from the hospital near the drilling sites signed a petition declaring their objection to the project (Briggle, 2015).

## RESEARCH QUESTIONS

1. How did campaign advocates from "Frack Free Denton" and "Denton Tax Payers for Strong Economy" construct fracking in general?
2. How did each of these groups challenge the claims-making activities and goals of their adversaries?

## METHODOLOGY

### Data

In this study, we selected grounded theory (Charmaz, 2008; Glaser, 1978; Strauss & Corbin, 1998) on the power struggle of two grassroots groups

(Frack Free Denton and Denton Tax Payers for a Strong Economy) over fracking in Denton since it is a dominant methodological approach for this research. Our goal is to employ grounded theory to look for a theory that is methodically linked with the fracking as a social problem. To be able to explain this phenomenon, the researchers collected data from in-depth interviews, newspaper articles, letters to the editor, and campaign advertisements. Since the election resulted in a fracking ban in Denton, we assume that anti-fracking grassroots group reached out to the citizens to generate local consciousness of constructing fracking as a social problem.

## Participant Protection

We received IRB approval from Texas Woman's University and we followed IRB standards and ethics. We recruited interview participants through snowball sampling alongside social media platforms (e.g., Facebook). If the interview was face-to-face, we provided an informational sheet to interviewee. If the participant agreed to the terms of the interview, we scheduled an interview, either at that time or at a later date, depending on the availability and convenience of the interviewee.

Participants felt more comfortable to meet in public spaces (e.g., local coffee shops) for an interview. In order to address participant's privacy, we surely selected a spot that was sufficiently far away from other individuals to ensure that interviews were not overheard. We asked each participants consent for the audio recording of the interview. We also informed them if they refuse to be recorded, we will take notes through interview. However, all participants consented for the audio recording. We recorded 20 interviews for this research. After the interviews, the researchers reiterated that both researchers and interviewees reviewed the consent form that he/she has agreed to be recorded. At the end of the interview, we asked the interviewees if they have any questions.

## Participants

The researcher interviewed ten participants from DTPSE and ten participants from FFD. All participants (10) from DTPSE are male. These include three mineral owners (leaders at DTPSE), two previous mayor, one gas company owner, two retired residents, one lawyer, and one oil and gas attorney. Four participants from FFD group were female. The participants included four leaders from FFD group one professor at the one of the local University, one retired professor, one nurse, one working for Earthworks employer, one documentary director, two vocalists, one sculpture artist, one local business owner, and one student from Socialist Student Association.

## Sampling

We interviewed with campaign supporters from Frack Free Denton (10) and Denton Tax Payers for a Strong Economy (10) Researchers conducted interviews in April 2015. The interviews are in-depth and semi-structured. Some of the questions are: When and how did you hear about fracking? What kind of venues do you/your movement use to reach out local citizens about Fracking? (Social media, newspapers, flyers, face-to-face meetings, TV, Radio, Letter Campaigns to Legislators, Door to Door, media, YouTube, Billboards) How do you frame fracking issue to persuade Denton citizens? How would you define your Volunteer activities, campaign responsibilities?

## Data Analysis

Back in the office, the audio recordings were transferred into transcript and subsequently erased from the recorder, all the relevant date was deleted and purged from the computer. The audio recordings of the interviews were transcribed. We were guided by the coding steps (Strauss & Corbin, 1998). Jotted notes in the data sheet were organized as field notes. All collected data · were coded into the three main categories. However, the constant comparison of data enlarged the focus of the analysis.

Coding started with open coding in order to code data according to preliminary characteristics. In this technique, we inspected our documentation created from interviews by concentrating on fracking, the claim making activities, and the goals of Frack Free Denton and Denton Tax Payers For Strong Economy. Axial coding followed the open coding by selecting key concepts leading to the research topic. Some of the initial codes that appeared from the open coding course were property rights or human rights, news advertisement, blog entries, websites, puppet show, and kids in action, all these codes generated to the category claims making activity. In order to select the key concepts, we examined all the evidences carefully and we organized the codes based on frequent themes. These themes developed key candidates for common categories that is linked to number of associated codes. Selective coding completed the coding. Mapping, as a third stage of the analysis of the organized data, helped the researchers see all collected data in a logical order (Strauss & Corbin, 1998).

## FINDINGS AND DISCUSSION

A review concerning the fracking controversy will be helpful to understand · how these grassroots groups (gas drilling companies, environmental entities, etc.) are grounded on their side of the controversy.

The claims-makers of this study are the interviewees of opposing groups, which are referred as the FFD and the DTPSE. The claims-making activities will be embedded in the aforementioned stages of development of a social problem. Due to the overlapping nature of the stages, they could not be specified individually in every case.

## Claims-Making Activities

Claims-making activities are drawn from the interviews. The Frack Free Denton group outnumbered the Denton Taxpayers for a Strong Economy group's activities. The FFD group recruited volunteers from local citizens to accomplish their goals during the campaign. On the other hand, the DTPSE worked with a private PR company in order to raise awareness during this stage.

## Awareness Stage-Frack Free Denton

In awareness stage, FFD organized several activities to raise awareness. The claims-making activities in this stage are as follows: canvassing, demonstration, information booth, websites, blog entries, Facebook, puppet shows, flash mob dance shows, media outlets (e.g., concert, documentary, YouTube), kids in action, yard sign, and billboards. In the policy determination stage, FFD engaged in activities to construct the ordinance about fracking to make fracking safe to Denton. The claims-making activities in this stage include letters to the editor, panels, city council meetings, and mic checks in the city council meetings. In the reform stage, FFD participated in activities to make pressure on elected officials and to inform the locals about the current development about fracking. The claims-making activities in this stage are as follows: phone calling officials and bus trip to Austin.

To raise social problem awareness, the FFD group used various venues to reach out to local residents and explain their cause during the awareness stage. These venues are canvassing, demonstration, websites, block entries, Facebook, yard sighs and billboards, puppet shows, flash mob dance shows, media outlets, and kids in action.

> Anti-Frack, Case 4: One of our first venues, well we had the signing party at Sweetwater Grill which is a place that's very popular in Denton. Then we went to the Mardi Gras party. We went to Churches, we had Gas Land movie showings, we set-up on the Square, just set-up. Puppet shows, music events, all kinds of things like that.

## Canvassing

The FFD group knocked on 90% of the doors in Denton with the help of volunteers. The leaders of FFD prepared literature to hand out. Volunteers were trained to use talking points during their canvassing. They also had maps with the exact location of houses they needed to visit. The volunteers were wearing "I live in Denton and We're not getting paid" stickers. In terms of awareness stage, one of the main approaches to convey their messages is canvassing. Canvassing is to inform neighbors concerning their claims and claims activities in local area by vising each resident's house.

> Anti-Frack, Case 6: . . . They would knock every single door on that block, talk to their neighborhoods, say, "Hi, I'm your Frack Free Denton. This is what fracking is. Here's some materials.

## Demonstration

The FFD group mobilized locals to protest HB40 rules restricting local control. In these demonstrations, volunteers carried posters and banners. Moreover, some volunteers presented their art works to show their stance against fracking.

> Anti-Frack, Case 7: . . . I tried running to bring that forward and that's just the way of doing it. My deal was my degree of activism would be I will walk. That's the key point. Everybody's got a place and they grab it and take it. As long as we diversify we have a better chance to conquer.

## Websites

Frack Free Denton owns an operational website: www.frackfreedenton.com . The claims and counter-claims are presented in this website. The website is frequently updated. The local websites are also intensely used to disseminate the awareness of claims and claims activities by FFD activist and volunteers. This venue is another important way to communicate and recruit local residents into their community fracking concern.

> Anti-Frack, Case 2: I'm on a web-based neighborhood communication site and during the campaign, it was very active site. A lot of people didn't want us talking about it, but we kept talking about it anyway, because that's part of the whole set-up and the communication site is that you're to discuss things that have an impact on your community.

It is noteworthy that one of the interviewees owns a website to inform people about fracking and the hits are around 2.5 million. The website address is www.texasharon.com.

## Blog Entries

Some of the FFD volunteers engaged in writing their reflections regarding the fracking issue in their blogs, sharing the photos and videos, and posting their blog entries on their Facebook and Twitter account to spread the word. To raise awareness, volunteers and activists used social media including opening blogs to have a unified local voice against fracking.

> Anti-Frack, Case 3: To me, social media is the big one. So I meant a blog. I used to have my own blog, Denton Drilling and then as the campaign heated up for the ban, I felt the need for us to speak more in unified voice and have sort of one platform so I switch my blogging over to Frack Free Denton and so I got lots of blogs automatically, yeah feel free to move stuff around.

## Facebook

The FFD group actively used Facebook to share information. More than 9500 people followed their pages. The FFD group announces events and posts updates in this account. To raise awareness, volunteers and activists employed Facebook to distribute their communication via postings of activity fliers, poll watching, providing memes, and their every upcoming social activity.

> Anti-Frack, Case 8: Did a lot of advance help with social media like sharing stories on Facebook and creating little memes and stuff and posting online Facebook. And then also just getting the word out by mouth and telling whoever would listen about the Frack Free Denton campaign. And then right-up at the election, we were doing a lot of the poll watching and I guess not really poll watching but you know, handing out fliers at the polls.

## Yard Signs and Billboards

The FFD group distributed yard signs to whoever wanted to display them. Moreover, FFD volunteers asked locals in Denton if they would like to have yard sign to show their support while doing canvassing.

## The Art of Fracktivism

*Puppet show:* The FFD group utilized the universal language of arts to reach out to the locals. The FFD group made their claims through a puppet show to reach out kids and families to promote awareness in the city of Denton.

> Anti-Frack, Case 5: . . . includes the families and otherwise and it sucks when families have to get somebody to take care of their kids so that they can go participate in politics. She described music as a universal language that can appeal to all ages and that puppets can appeal to all ages. It's good because they educate everyone, the opposite of that would be things that are locked in

to really academic jargon, buried in the papers. You can't expect everybody in Denton to read a dissertation, but you could get them all to watch a puppet show.

## Flash Mob Dance Show

To promote awareness and to facilitate a meaningful debate on the fracking issue, two dance professors from Texas Woman University coordinated a flash mob dance to demonstrate: banning fracking. Anti-Frack, Case 8: Basically, like an improvisational dance routine that's some of the TWU dance professors. They were the ones that coordinated that.

> Anti-Frack, Case 6: We had on the Flash Mobs. For that, we had R. but we also had there is this other Lady who did the dance. Her name is SG. SG, she is also a professional TWU. She's a dance professor.

## Media Outlets: Music, Documentary, YouTube

FFD group actively utilized the media outlets. They organized a concert for fund raising and also to raise social problem consciousness. Brave Combo, a Grammy award winner and nationally recognized band that is located in Denton, composed a song about fracking called No Fracking Way.

> Anti-Frack, Case 2: Brave Combo is a local band who had been together for 30 year and they started as a garage band, we call them. They're just a bunch of kids playing together. And then, won two Grammy's and they are like a V-band from Denton.

One of the volunteers of FFD group prepared the frackettes video, hits of 21,000, and a famous environmental activist, Erin Brockovich, shared it on her Facebook page. The video can be seen at: https://www.youtube.com/watch?v=MD5r8WGYAug.

### *Kids in Action*

FFD organized family friendly events to reach out to children of the city to inform local families in the awareness stage of claims making activities. One of the activities was painting yard signs and drawing pictures about fracking.

Anti-Frack, Case 6: A drawing made of one of the children from the neighborhood was one of the first thing we ever put on the Frack Free Denton website. And it said, on one side it was bright and cherry and it said, Denton without fracking, and the other side was dark, the home was cracked and it said, Denton with fracking. Those kind of ideas like that, I think, they do, they tell of it's I don't know and yet it is very important to see the way children see it.

## Policy Determination-Frack Free Denton

During the policy determination stage, focusing on what should be done and proposing solutions to the social problem, the FFD group used several venues to reach out to the public figures, authorities, community leaders, and local politicians in order to shape the ordinance of the fracking regulation and to discuss the petition. A quote from one of the interviews summarizes the whole stage pretty well.

> Anti-Frack Case 3: They were planning in making a park there and somebody told us about the gas wells are going in and no idea what they are talking about so I went down to a protest at city hall because we had heard about it and it was mostly just the educate ourselves and talk to people. And they were protesting those wells and I started to learn about on my own a bit what fracking is this is before gas lane came out. All right so I think it was all there. It wasn't on the national consciousness here and so it was about a year later that KR who was newly like the city council member came here and approaches me and others at the former center for the studies and the disciplinary, which used to exist here and the UNT but they defunded it recently. And he said look, we are revising our ordinance. This is an interdisciplinary issue you know it's engineering, its science, its law, its ethics, it's all of the stuff mixed together. Would you like to form a grass roots shadow advisory commission, that's the way he put it, unofficial but the city had formed an official task force which they have majority industry members on it and Kevin said, we need more of a counter balance from citizen prospective. So that's how I got really officially involved. I mean our group was unofficial by grass root really involved was through him coming to me and that was a birth of what we called then the Denton's, they called Drilling Advisory Group DAG.

## Protest during City Council Meeting

In the course of the policy determination stage, another emphasis is on "what should be done" to deliver the message to the authorities and local politicians in order to form the ordinance of the fracking regulation. During one of the city council meetings, the activists protested the current regulation of fracking. The protesters asserted that the fracking in Denton polluted our environment.

> Anti-Frack, Case 5: . . . Media is the one thing that I can really do but prior to that, my role was not very big but I would participate in the demonstrations, I would show up city council meetings, there was a YouTube video that I made a long time ago when Occupy Denton, mic checked the city council. I don't know if you are familiar with that strategy. You have to have permission to get the microphone but if you all work together, you can overpower that and you could get your message out so even though the city council has the microphone after they made their decision to drag their feet or whatever they do, somebody yells mic check and that means get ready we're about to do this, so

mic check and then everybody else yells mic check and then one person shouts basically a message.

## Social Media Campaign

The FFD group organized social media campaigns for local control. The volunteers created hash tags such as #ProtectLocalControl. They encouraged their followers to tweet and share a photo of a sign stating why protecting local control is important. They urged their followers to reach out to the local representatives and the local media sources to help spread the word. They continued to create hash tags such as #ProtectLocalControl, #StandWithDenton, #StopHB40, #Stop1165, #DefendDemocracy, #frackfreedenton.

## Raising Fracking Consciousness for Kids

One of the oil and gas companies made a coloring book about fracking for kids. They made claims about how fracking is safe for our environment. The oil company supported DTPSE. The sample screenshot of the paint book is below.

## Mails

One of the campaign strategies was to send informative mail to local residents. The ads in the Denton Record Chronicle and the mailing materials were similar. Since the mailing brochures are expensive, FFD advocates did not use mailing as a strategy to reach out locals. Instead, they were knocking doors to hand their informational brochures to the locals.

Findings related claims-making activities show that the two contending groups engaged in throughout the stages. In accordance with the claims, the claims-making activities were higher in numbers for the FFD. The table shows that the groups have both engaged in some activities; however, the quantities were different for each group. For example, both groups have attended city council meetings. However, DTPSE rarely attended the public hearings, while FFD has been present most of the time. FFD volunteers took the platform and mentioned their concerns repeatedly.

## CONCLUSION

The findings represent an exploration of how anti-fracking and pro-fracking groups engaged in a power struggle over fracking in Denton, TX. Lemert's findings show that public officials received complaints in Oakland against trailer camp problems from only individual awareness level (Lemert, 1951). However, in terms of an awareness of social problem, there was no organized public opinion neither generated nor channeled through the media (Lemert,

1951). Although Lemert studied the natural history concept from the viewpoint of its applicability to the trailer camp problem in California communities by following the theoretical scheme stated by Fuller and Myers, this research was the first to contribute to filling part of the gap in information on the propaganda activities of two rival groups as claims-makers. The major claims are identified with the analyses of the in-depth interviews, which explored the claims of each group in greater detail. The major claims are discussed through the stages of awareness, policy determination, and reform.

FFD made a greater number of claims. Second, FFD efficiently rebutted DTPSE's claims with its counter claims. Third, FFD did more numerous local volunteers than DTPSE, which mobilized paid volunteers for its cause. Fourth, FFD had a greater number of claims-making activities that DTPSE. Fifth, the FFD claims-making activities were enhanced by artistic and creative events such as a puppet show, concert, and sculpture. DTPSE used traditional venues of claims-making activities. Sixth, FFD's grassroots and local lobbying overshadowed the DTPSE's effort. Seventh, FFD utilized DRC more effectively than DTPSE. Despite the local newspaper's editorial support for fracking, FFD was able to capture the DRC's audience without advertising and through activities such as letters to the editor, press releases, guest columns and arranging newsworthy local events that were covered by DRC.

We contribute a more comprehensive understanding of value-conflict and social construction theories. We also believe that this study will guide community leaders over the fracking issue that will remain ongoing for many years.

## BIBLIOGRAPHY

Best, J. (2017). *Images of issues: Typifying contemporary social problems.* London, UK: Aldine Transaction.

Brasier, K. J., Filteau, M. R., McLaughlin, D. K., Jacquet, R. J., Stedman, C., Kelsey, T. W., & Goetz, J. S. (2011). Residents' perceptions of community and environmental impacts from development of natural gas in the Marcellus shale: A comparison of Pennsylvania and New York cases. *Journal of Rural Social Sciences, 26*(1), 32–61.

Briggle, A. (2015). *A field philosopher's guide to fracking: How one Texas town stood up to big oil and gas.* New York City, NY: Liveright.

Charmaz, K. (2008). Grounded theory. In J. A. Smith (Ed.), *Qualitative psychology: A practical guide to research methods* (pp. 81–110). Los Angeles, CA: Sage Publications.

Environmental Protection Agency (EPA). (2018). The process of unconventional natural gas production. *Hydraulic fracturing.* Retrieved from https://www.epa.gov/uog/process-unconventional-natural-gas-production .

Fuller, R. C., & Myers, R. R. (1941). The natural history of a social problem. *American Sociological Review, 6*(3), 320–329.

Glaser, E. G. (1978). *Advances in the methodology of grounded theory: Theoretical sensitivity.* Mill Valley, CA: Sociology Press.

Gullion, J. S. (2015). *Fracking the neighborhood: Reluctant activists and natural gas drilling.* Cambridge, MA: MIT Press.

Jones, P, Hillier, D., & Comfort, D. (2013). Fracking and public relations: Rehearsing the arguments and making the case. *Journal of Public Affairs, 13*(4), 384–390.

Julios, C. (2015). *Forced marriage and honour killings in Britain: Private lives, community crimes and public policy perspectives.* La Vergne, TN: Lightning Source Inc.

Lemert, M. E. (1951). Is there a natural history of social problems? *American Sociological Review, 16*(2), 217–223.

Linton, D. S. (1991). *Who has the youth, has the future. The campaign to save young workers in imperial Germany.* Cambridge, UK: Cambridge University Press.

Rabe, B. G., & Borick, C. (2013). Conventional politics for unconventional drilling? Lessons from Pennsylvania's early move into fracking policy development. *Review of Policy Research, 30*(3), 321–340.

Strauss, A., & Corbin, M. C. (1998). *Basics of qualitative research: Grounded theory procedures and techniques.* Thousand Oaks, CA, US: Sage Publications.

Witter, R. Z., McKenzie, L., Stinson, E. K., Scott, K., Lee, S. N., & Adgate, J. (2013). The use of health impact assessment for a community undergoing natural gas development. *American Journal of Public Health, 103*(6), 1002–1010.

## Chapter Three

# Theories of Environmental Social Problems

*A Synthesis of News Media Play in the Power Struggles*

By Sebahattin Ziyanak and Mehmet Soyer

During the election on November 4th 2014, Denton city voted to ban fracking. Approximately 59 percent of voters supported the ordinance that read as follow: "Shall an ordinance be enacted prohibiting, within the corporate limits of the city of Denton, Texas, hydraulic fracturing, a well stimulation process involving the use of water, sand and/or chemical additives pumped under high pressure to fracture subsurface non-porous rock formations such as shale to improve the flow of natural gas, oil, or other hydrocarbons into the well, with subsequent high rate, extended flow back to expel fracture fluids and solids."[1]

This chapter examines an in depth explanation of objective and subjective sociology of a social problem. "Value conflict theory" and "Social construction" theories are in the area of sociology of social problem. These theories constitute theoretical framework for this study. In this chapter, we propose a synthesis of these theories for this research. Furthermore, we review the literature related to the concerning news media. Finally, this chapter elaborates the findings about research question of Hydraulic Fracturing (a.k.a. fracking).

## OBJECTIVE VERSUS SUBJECTIVE
## SOCIOLOGY OF SOCIAL PROBLEMS

Social problems have been defined by Bassis et al. (1982) as specific social conditions that cause harm to the society as a whole or directly to the individuals in it. This definition uses objective criteria to evaluate individual social cases as genuine social problems, based on their potential or actual harm to society and individuals. An example of such a perspective is the social pathology theory.

There are two problems with the objectivist definitions. They tend to obscure the potential subjective aspects of social problems. For example, the majority of scientists have a consensus regarding the destructive role of certain manufactured chemicals in corroding the ozone layer, which in turn directly contributes to a rise of skin cancer incidents, agriculture crop damage, and other undesirable consequences. Thus, at first glance, this phenomenon might easily be classified as a social problem under objectivist terms. However, this consensus first had to be reached by scientists in the past through socio-cultural consensus-making processes. Scientists had to first discover ozone depletion and its causes, then convince their colleagues and the general public of the seriousness of the problem. In addition, this process had to be carried out while the chemical industry fought to deny such claims, politicians and the press had their own incentives to bring the issue to the public attention. All of the above indicate the subjective aspect of social problems. In the hypothetical cases where scientists did not succeed in making such discoveries, or politicians and press did not push to bring these factors in to the public light, ozone depletion might not have been recognized as an official and objective issue. However, this lack of objective recognition would not have altered the objective truth about its severe side effects, such as skin cancer, etc., continuing to occur on an increasing scale.

The second flaw of using objectivist definitions determining what objective criteria to use in this defining process. Even in the case that a consensus is reached regarding what issues are considered as actual social problems, it is unlikely that another consensus will be reached regarding the criteria used to define their objectivity. Aside from the general fact that they all have negative effects on society, it is quite a difficult task to generate other underlying causes and criteria that can be considered objective and universal.

Due to such evident limitations of the objectivist method, some sociologists have turned to developing subjective methods to classify social problems. Spector and Kitsuse (2017) have identified this method, sometimes called the subjectivist approach, as social constructionism. This theory is based on the belief that social problems are produced or 'constructed' in the eyes of society and individuals via exposure and participation in social activities. For constructionism, it is not the mere existence of an issue, but rather its induction into social life through media coverage, public demonstrations, and legal discussions in courts and political bodies that constructs and establishes it as a social problem. The key event in this process is the contending groups activities. This is what Kitsuse and Spector (2017) call claimsmaking.

Objective sociologists consider issues as social problems based on their effects, whereas constructionists consider their classification based on their claimsmaking and how concretely they are recognized by society as social problems. Objectivists recognize that subjectivity and recognition of social conditions is a relevant factor, but do not take them into consideration in classifying the conditions as social problems or not. For objectivists, the focus is on the conditions and the adverse byproducts of social problem. On the contrary, subjectivists do not put emphasis on the conditions themselves, but rather focus on how well society recognizes these conditions and how well it accepts them as reasons for concern.

This difference in defining social problems does not imply that constructionists are empathetic toward society's well being. The difference in definition applies to the methodology in analyzing social problems and their construction. Constructionists will focus on the claimsmaking aspects and attempt to analyze how social conditions lying beneath the problems are constructed into recognized issues for society.

## VALUE-CONFLICT THEORY:
## NATURAL HISTORY APPROACH

Richard Fuller and Richard Myers (1941) declared in their remarkable article, The Natural History of a Social Problem, "It is our thesis that every social problem has a natural history and that the natural history approach is a promising conceptual framework within which to study specific social problems (Fuller & Myers, 1941, p. 320)." Fuller and Myers provided developmental stages of a social problem which will be discussed on in the following parts of this chapter. However, first, it is urgent to explain how certain conditions become social problems.

The first proposition is that social problems are viewed as a deviation from the norm. The society considers the deviation as a threat; therefore, deviance does have an objective condition and a subjective definition. From this point of view, objective conditions do not constitute social problems. Objective condition is a verifiable situation, which means it can be measurable and observable by experts. For example, air pollution is a verifiable situation and can be measured and reported to the public. However, different groups in the society construct the reasons for the consequences air pollution differently. Subjective definition is based on the awareness of a condition as a threat to the societal norms. Therefore, different societal phenomena can give different meanings to the deviations and define them accordingly, which brings us to the next point.

The second proposition emphasizes the fact that the existence of an objective condition is not sufficient for phenomena to become a social problem. If the people in the society do not define a condition as a problem, it will not be treated as a social problem. Because, social problems are what people consider them to be. The conditions can be a problem for others in the society, but not necessarily for this specific group not so long as that they define the condition as a problem. For example, air pollution can be stated as a social problem for some specific environmental groups in a specific area, but the society in that specific area may not define the condition as a social problem. Once, the society is convinced that the condition exists, it becomes a social problem. Overall, the objective condition holds importance in terms of defining a social problem, but not sufficient if subjective definitions are not in play.

The third proposition states that cultural values have a direct causal link to the social problem development through the objective condition. As a society meets on a common ground when it comes to cultural values and beliefs, individuals are able to maintain social institutions and order. The fourth proposition builds on sustainability of the cultural values and beliefs. Fuller and Myers (1941) state that cultural values can turn into obstacles every now then. The idea behind this reasoning is that, since society is focused on sustaining these values and beliefs, any deviation from the norm is a potential threat to the social institution. Therefore, individuals become reluctant to declare any amelioration to the situation. In such cases, the solution will be considered as a violation of the mores, and abandonment of beliefs. As a result, the cultural values become an obstacle in the phase of transformation.

Dual conflict of values follows the previous proposition. Dual conflict arises when a party considers the condition as a threat and the other party doesn't. The discrepancy results in the opposing beliefs of the parties involved. There are two different versions of the dual conflict. First one occurs when the disagreement is over the deviation from fundamental values, and the second one happens when there is an agreement over the existence of the threat but the ways to resolve the issue is a matter of conflict. Therefore, social problems exist and are sustained so long as the individuals do not share the same values and beliefs.

Finally, in the field of sociology, we do not study the objective condition alone, but also subjective definitions. A condition can exist in a society in which different parties may have different values, different definitions and different solutions to the condition. Therefore, value judgments of the involved people become a key part of the development of a social problem.

Fuller and Myers (1941) propose this framework as a useful tool to study social problems. They also refer to natural history of the rise of the social problem. The framework underlines the shared characteristics of all social problems. Built on the discussed propositions, all social problems go through a common sequence: awareness, policy determination, and reform. The following section discusses these stages in details.

## SOCIAL CONSTRUCTION THEORY

In the 1970s, Spector and Kitsuse (2017) were the first sociologists to define social problems as "the activities of individuals or groups making assertions of grievances and claims with respect to some putative conditions" (p.75). Social constructionist perspective considers social problems without regard to their physical properties. From a sociological perspective, people make claims about "alleged conditions." Therefore, social problems are generated through claimsmaking activities (Best, 2017). The claimsmakers construct the social problem (claim) through claimsmaking activities that aim at shaping the public perception. Best (2017) describes social problems as "emerging through the efforts of claimsmakers who bring issues to public attention. By typifying a problem and characterizing it as a particular sort, claimsmakers can shape policymaking and public response to the problem." Therefore, social constructionism requires focusing on the claims, the claimsmakers, and the claimsmaking process.

Constructionism claims that society's perceptions of social problems are social constructs. However, the role of claimsmaking in creating these social constructs goes beyond just raising awareness about that social problem. Claimsmakers go as far as to influence society's perception of what the problem actually is. Any social condition has a potential to be a social problem, and thus, can be the subject matter of claimsmaking. How the claimsmakers go about in doing that will define society's perceptions, whether it is through their choices of which aspects of the issue they will focus on or even how they chose to name the problem. This process of characterizing and classifying a social problem by the claimsmaker is known as typification. A significant example of typification is the claimsmaker choosing the orientation of approach to the social problem. By giving orientation, the claimsmaker suggests that the social problem be considered a certain type of problem, such as an economic, political, or moral one. Thus, different orientations put different spins on social problems, leading to various interpretations of the issue along with its suggested root causes and potential solutions (Best, 2017). Claimsmakers can target different audiences with different goals. Claimsmaking can be tailored to raise public awareness, target smaller groups that are being directly affected, aim to dissuade the ones responsible for the problem, or possibly influence the lawmakers or other points of authority and regulation. This can be carried out through various methods, such as direct communication. However, more effective and wide-reaching modes of communication, such as mass and social media, are preferred. In cases where they see fit, the press will also propagate the claimsmakings and assist with the construction of social problems.

At its core, claimsmaking is an attempt to persuasion. Claimsmakers aim to convince their audience that their topic at hand is an actual social problem and/or that the problem should be solved through his proposed solutions. Thus, the success of the claimsmaker is measured by whether his audience was convinced by his claims. Furthermore, analyzing these claims through regular rhetorical methods can give insight for their legitimacy and typifying effects. In cases when social problems are still very much open to debate, there can be multiple claims making cases for the same problem that compete with each other to gain legitimacy. This is expected since there can potentially be numerous varying typifications for the same social problem. Audiences will usually disregard most claims in such cases of ambiguity until certain claim gains dominance through a persuasive typification of the problem (Best, 2017).

## NEWS MEDIA

Media is a significant arena of claims-makers for power struggle. In this case, the battle was between pro-drillers for gas-drilling companies pointing at the economic advantages and anti-drillers pointing at environmental health concerns. While gas drilling companies have economic power, their media-source relations do not always determine what the media coverage is about natural gas development, particularly at a time when varying political and economic groups, as well as NGOs, have their own agendas, and lobby their own perspectives as the most legitimate and immediate concern (MacDonald, 2003, p. 39). Media industries "collect information, make decisions about the selection and presentation of programs and to a certain extent control the entry of topics, contributions, and authors into the mass-media dominated public sphere" (Habermas, 1996, p. 376). Since the media has the potential to inform individuals regarding environmental issues, and supplements their knowledge about it (Smith, 2002), the grassroots groups see media as a venue to reach out to the public.

### Fracking

It should be noted that the focus of this study is not on whether fracking causes environmental health issues or results in economic benefits. Rather, the focus is on how grassroots groups engage in power struggles over fracking through a Denton Record Chronicle (News Coverage, Campaign Advertisement, Letters to the Editor) and Campaign Volunteers. However, the literature review of fracking controversy is helpful to understand how these grassroots groups (gas drilling companies, environmental entities, etc.) are grounded on their side of the controversy.

To be able to grasp the fracking debate, the procedure of fracking needs to be comprehended. Fracking (hydraulic fracturing) is one way of extracting natural gas from shale rock. This process contains "clearing land for well pads; construction of access roads and ancillary infrastructure (e.g., pipelines, compressor stations); transporting and processing fossil fuels extracted; transporting millions of gallons of water and wastewater for treatment/disposal; and bringing large (and often transient) populations to a community" (Boudet et al., 2014, p. 2).

As explained in the chapter I, merging theoretical framework of value-conflict theory (Fuller & Myers, 1941) and social construction of social problem (Best, 2017) are necessary to understand Soyer-Ziyanak's stages of a social problem model. This model suggests that it is necessary to analyze the stages of awareness, policy determination, and reform concerning fracking by observing their claims, claims-makers, and claims-making activities.

## FINDINGS AND DISCUSSION

### Othering One Another Us vs Them

The exclusionary language use of both groups is remarkable. There were a lot of we, us and they, them words used to indicate the opposing group. Some examples are provided below to see what type of othering language is used in the interviews.

> Pro-Frack, Case 20: "Well the Chamber, very intentionally, Chamber leadership very intentionally decided not to go to either polarized extreme. We disused it and decided that that is not what we would do. We would not come out and say absolutely the ban is bad let's do away with the ban. We said no. The ban is bad but we must reasonably regulate. And so-so we took a middle ground and then we get attacked which tends to be what happens to middle ground. When you're being reasonable sometimes. The emotions of the extremes attack anyone in the middle. It's just what happens."

### Eco-Terrorists

FFD group mentioned that they have been called various names such as eco-terrorist by DTPSE group. They also claimed that their opponents either made a list of their names or pre-intended doing so. I'll list these "names" and provide examples below.

Anti-Frack, Case 9: "I've been even called an "eco-terrorist" so many times that now many times someone says the word "terrorist" that I might listen and giggle. Like, there was someone in Mansfield who said that he needed a bodyguard because the eco-terrorist were going to get him.""

## Corruption

Corruption is another claim that has been provided several times in the interviews. Both sides state that the other group will engage in dirty politics in the process of passing local control bills and other legislations. Some examples are below.

Anti-Frack, Case 1: "My only concern about the lawsuits is that our Texas courts, the judges are elected. So because they are elected, that means the industry gives them lots of campaign contributions. You would hate to think that that would buy a vote but you can't absolutely say it doesn't. I feel the lawsuits—we knew it was gonna happen. We were ready for it to happen. I wish it wouldn't because it's taking a lot of time and money away. But it is what it is. That's democracy. I'm gonna sound very jaded and very cynical but I think the industry—as in Austin right now trying to buy our elected officials to get them to pass laws to keep this from happening in the future and to overturn the ban."

## Media Outlets: Music, Documentary, and YouTube

Words and music is composed by Little Jack Melody of Brave Combo. It is called "Song: No Fracking Way!" Their main message is about how fracking has affected happiness in Denton, Texas.

"But tomorrow? Tell your friends, No Fracking Way!"

"Tell the suits, No Fracking Way!"

One of the volunteers of FFD group prepared the frackettes video, hits of 19,000, and a famous environmental activist, Erin Brockovich, shared it on her Facebook page.

## Yard Sign and Billboards

FFD group distributed the yard sign for whoever wants. Moreover, FFD volunteers asked locals in Denton if they would like to have yard sign to show their supports while doing canvassing.

The leaders of DTPSE indicated that they distributed yard signs to the local residents who wanted to show their supports. They were reached via phone or website to have yard signs. The billboards functioned as media outlets to reach out to more local residents. Even though the participants from both groups mentioned that wording "responsible drilling" is smart way to influence the locals. Some supporters from DTPSE stated that to write vote no drilling ban is somewhat confusing since people vote for or against the ban.

## Awareness—Denton Tax Payers for Strong Economy

DTPSE was working with a PR Company to do their campaign during the election. The company had never lost any election in previous jobs. The second author requested to conduct an interview with the individual at PR Company to discuss the campaign strategy of DTPSE. However, the individual for the PR company did not grant an interview.

## Reform—Denton Tax Payers for Strong Economy

The leaders of DTPSE stated that they are not doing any effort in reform stage. However, the oil and gas corps took an action in this stage to lobby at Austin to overturn the fracking ban in Denton.

Denton Record Chronicle is a prominent local newspaper in Denton. Each group has used DRC as a venue to reach out and inform the local citizens. Even though both groups used DRC as a tool, they also voiced reservations. FFD group reached out to DRC if they had any events. They called or emailed or they do issued press release to DRC. However, DTPSE group's perspective on reaching out differs from FFD group. One interviewee from DTPSE, PRO-Frack, Case 13, states, "X is the reporter. X is the one that should reach out in my opinion. If X is needing a story, then X should reach out to both sides, not just one side. X never reached out to our side." However, one of the reporters from DRC pointed out that when he/she calls a representative from Denton Taxpayers for a Strong Economy, DTPSE didn't answer the phone or did not return the call for a comment.

## Letters to the Editor

One of the strategies used during the campaign was sending Letters to the Editor. Both groups took advantages of this section of the newspaper at Denton Record Chronicle to make their voice heard by public. These letters raised attention for readers and the readers made controversial commentaries about the issue. Sometimes the readers had heated debates with their commentaries. FFD was more organized in using this tool than DTPSE. The supporters of FFD outnumbered DTPSE in terms of sending letters to the editor. As a result, both groups have been found to use letters to the editor to convey their messages and to engage in power struggle over fracking. Both groups constructed the issue of fracking in different ways in favorable of their positions. Since a social problem is constructed through conflicting values and interests, each group emphasizes the cultural values that are important to them. Therefore, the construction of the problem through local

news media involves these two groups as claims-makers clashing over the
issue of fracking.

Denton Taxpayers for Strong Economy strongly asserts that fracking ban
is unconstitutional under the law. They believe that the case is sure to be
overturned from legislation eventually. An excerpt from the DRC's letters to
the editor is presented below as an example.

> . . . As Dr. Ed Ireland pointed out in the DRC opinion column on May 14,
> fracking bans across the country are costing taxpayers. Cities and counties are
> hiring lawyers to defend the policies, which are being overturned left and right
> as illegal or even unconstitutional. We Denton residents deserve to know how
> much it has cost us to defend our ban . . . (Sweet, 2015).

DRC archives revealed that the trend of the power struggle is continuous.
If FFD supporters have claims about fracking, DTPSE supporters provide
counter arguments and it goes back and forth between both parties. Here are
some excerpts exemplifying the debates.

> Both Frank Mayhew's "Attitude adjustment" [DRC, 3-17] and Ed Soph's
> "Stand up for us" [DRC, 2-17] reverberate the common theme that the "People
> have spoken" through the Denton fracking ban, implying the vote represents a
> mandate. . . . Frack Free Denton described the neighborhoods of Vintage and
> the Meadows at Hickory Creek (located in Precinct 4003) as the top reason to
> ban fracking in Denton because wells were drilled and fracked within 200 feet
> from homes there. The election results showed a different consensus. Precinct
> 4003 voted against the ban by a 2-1 margin. The 1,794 votes against the ban as
> opposed to 952 votes for the ban represented the widest margin in the city.
> Conversely, the largest support for the ban came from Precinct 4007, which
> includes part of the UNT campus. There, 1,452 voters supported the ban while
> only 347 opposed it. Wendy Davis won Precinct 4007 by a 3-1 margin but
> Gov. Greg Abbott carried Denton County by 30 points. This is a mandate?
> Referendums should not be used to destroy representative government but to
> correct it when it becomes misrepresentative. Mayhew's and Soph's challenge
> that elected officials march in lock step with Frack Free Denton is akin to a
> tyrannical direct democracy, not a representative republic (Lawson, 2015).

Another argument that has been highlighted in DRC is whether fracking is
safe or unsafe. DTPSE's argument is that fracking is safe, whereas FFD
continuously refer to fracking as unsafe. On the one hand, DTPSE emphasizes that whatever evidence FFD presents an issue is unscientific, and highly
partisan. FFD indicates DTPSE focuses on the economic benefits (minor) of
fracking, at the expense of harmful consequences of the fracking.

> Denton is facing a big decision this November to decide whether or not to pass
> a ban on hydraulic fracturing within city limits. This is not a partisan issue.
> This is not a conservative or liberal issue. At its heart, this is an issue of health

and safety for our fellow citizens. Hydraulic fracturing of wells near neighborhoods is dangerous and risky, so that's why this November I'll be voting to pass the ban (Wicks, 2014).

DTPSE supporters state that FFD group use the fear factor to manipulate the truth about the natural gas drilling. DTPSE supporters mentioned in the letters the local residents should know the facts, rather than appealing to false manipulative information. Moreover, based on the truth about fracking, they encourage the local residents to vote no on the ban. One of the examples is:

> A Halloween scare tactic from the Frack Free Denton mailer alleges that fracking is the reason Denton has the most unhealthy air in Texas. The facts refute this myth: Since 2000, ozone levels have fallen while the number of gas wells has increased in the entire 10-county area of North Texas. A study at Southern Methodist University concluded that there is no clear relationship between Barnett Shale natural gas production activities and the highest average ozone levels. Most Barnett Shale natural gas wells produce dry gas, meaning that they produce no other liquids. Road and non-road vehicles produce more than 75 percent of the unhealthy emissions, and Denton's air also is affected by coal and other pollutants coming from points south by our prevailing winds. . . . Base your vote on facts, not false scare tactics. Vote no on the ban (Edmondson, 2014).

DTPSE stated that the intention of FFD is actually to ban natural gas drilling activities. They indicated that the industry would not drill the wells if they were not able to use hydraulically fracture technic, extracting significant amounts of natural gas. Along with economic benefits of fracking in Denton, DTPSE supporters also stated the fracking ban in Denton will cost millions of dollars to local residents. Moreover, the mineral owners will sue the city. Fracking issue became a power struggle ground for city council member candidates. To endorse one of the council member candidates, FFD group was urged in letters to the editor to recruit more people during city council member elections.

> Denton needs Kathleen Wazny to represent District 3 on the Denton City Council . . . Kathleen was one of the pioneers in the fight to restrict fracking activities within the city limits of Denton. . . . (Morris, 2015)

To ban fracking in Denton by local residents' votes brought a lot of attention from all around the country. Indeed it was frequently mentioned that this vote "put Denton on the map" Both groups indicated that the oil and gas corporations are very powerful organizations in Texas. Then, these corporations would lobby in Austin to change the proposed ban. Recently, the counter claims-making activities are to culminate in passing HB40 bill, which restricts local control over industrial development. Both groups generated claims on HB40 and struggled over HB40 bill. FFD used HB40 bills to exert pressure over elected officials.

## News Advertisements

The newspaper advertisements are also the claims-making activities to make claims about the fracking issue. FFD sees fracking as social problem. The major claims of FFD group emphasized in the local newspaper ads. In the ads, FFD stressed the harmful effects of fracking on Denton's children and the minor economic benefits of fracking on Denton's schools. They highlighted that the mineral owners are the one who makes the money from fracking. Moreover, FFD highlighted the rights of Denton residents since they deserve clean air, safe environment, and more valuable homes. Both groups accused each other of using scare tactics and manipulation of facts.

## Denton Tax Payers for Strong Economy-Ads

Interviewees from both group indicated using responsible drilling in the ads is good strategy to recruit local residents. DTPSE highlighted the economic benefits of fracking. As FFD emphasized on the future of Denton's children, DTPSE also highlighted how fracking will benefit the schools, the universities, and the parks through taxes. One of the criticisms of the FFD group was that DTPSE group did not use the local kids in their ads.

> Anti-Frack, Case 1: "They bought stock pools from Germany. Like the little girl in the swing, swinging, looking so happy, if you ban fracking, if you ban drilling then that poor kid won't have any money in her school. So what did we do? We looked it up, it's a stock photo from Germany, and we said, "Hell yeah, she looks happy. There is no Fracking in Germany." That's how stupid they were."

In the ads, they underlined the fact that the ban will cost millions of dollars to the city and the local residents. In the ads, the testimonies of the local leaders (previous mayor, previous TWU president, e.t.c) and the local organizations (Chamber of Commerce, State Fair) have been covered.

## News Coverage

Both Frack Free Denton and Denton Taxpayers for Strong Economy have used a local newspaper, Denton Record Chronicle, as a venue to reach out to local citizens. Since local newspapers focus on local problems, both groups published advertisements, reached to the editor for commentaries, and have been cited several times in the news coverage. However, the researcher's goal is not to focus on analyzing how journalists frame the news coverage regarding fracking. The focus will be on the important process of the claims and the claims-making activities of both groups. These groups and their news need to attract reporters to find a place in the coverage since the reporters in editorial meeting are deciding which coverage will take place. Moreover, sometimes the groups reach out to the local newspaper for a press release or reporters do reach out them.

## CONCLUSION

The claims, claims-making activities, and claims-makers of both groups were covered in the local newspaper in great detail. The coverage also includes some details of each campaigns fundraising activities. Some local business concerns have shown support for various claims-makers on the fracking issue. For example, some local coffee shops and restaurants were welcomed FFD meetings. Also, even though Denton Chamber of Commerce encouraged local residents to vote no on the ban, they stated that they did not authorize their testimony to be used in campaign materials produced by DTPSE.

Some local organization declared their side on fracking issue for both side. Even if Denton Chamber of Commerce encouraged local residents to vote no on the ban, they stated that they did not endorse to use their testimony as campaign flyer by DTPSE. Since the FFD group has made effective use of visual arts, sculpture, painting as well as music to the locals, Denton Record Chronicle also covered some claims-making activities of FFD group. It also covered claims-making activities even though they were fewer in numbers. Local news reports illustrate this fact. The oil and gas corps mostly supporting Denton Taxpayers find place in the news coverage to express their opinion about the fracking ban. The editorial board of DRC declared its position against fracking ban for particular reasons as explained in their editorial letter before the election. The president of FFD group published an open letter in DRC in order to exert pressure on elected officials and to let people know that the state representative acted against local residents' perspective on fracking ban.

## NOTE

1. http://www.cityofdenton.com/departments-services/general-election

## BIBLIOGRAPHY

Bassis, M. S., Gelles, R. J., & Levine, A. (1982). Social problems. New York, NY: Harcourt Brace Jovanovich Inc.

Best, J. (2017). *Images of issues: Typifying contemporary social problems.* London, UK: Aldine Transaction.

Boudet, H., Clarke, C., Bugden, D., Maibach, E., Roser-Renouf, C., & Leiserowitz, A. (2014). Fracking controversy and communication: Using national survey data to understand public perceptions of hydraulic fracturing. *Energy Policy, 65,* 57–67.

Edmondson, D. (2014, October 14). Act like adults. *Denton Record Chronicle.* Retrieved from http://www.dentonrc.com/opinion/letters-headlines/20141014-letters-to-the-editor-oct.-14.ece.

Fuller, R. C., & Myers, R. R. (1941). The natural history of a social problem. *American Sociological Review, 6*(3), 320–329.

Habermas, J. (2015). *Between facts and norms: Contributions to a discourse theory of law and democracy.* New York, NY: John Wiley & Sons.

Lawson, B. (2015, March 30). What mandate? *Denton Record Chronicle.* Retrieved from http://www.dentonrc.com/opinion/letters-headlines/20150330-letters-tothe-editor-march-30.ece.

Macdonald, M. (2003). *Exploring media discourse.* London, UK: Oxford University Press.

Morris, E.T., (2015, April 28). Kathleen Wazny best choice. *Denton Record Chronicle.* Retrieved from http://www.dentonrc.com/opinion/letters-headlines/20150428-letters-to-the-editor-april-28.ece.

Smith, E. R. (2001). *Energy, the environment, and public opinion.* Lanham, MD: Rowman & Littlefield Publishers.

Spector, M., & Kitsue, J. (2017). *Constructing social problems.* New York, NY: Routledge.

Sweet, P. (2015, May 20). Taxpayers deserve to know cost. *Denton Record Chronicle.* Retrieved from http://www.dentonrc.com/opinion/letters-headlines/20150520- letters-to-the-editor-may-21.ece.

Wicks, H. (2015, Oct 30). Pass The Ban, Retrieved from http://www.dentonrc.com/opinion/letters-headlines/20141030-letters-tothe-editor-oct.-30.ece)

*Chapter Four*

# Public Health and Environmental Concerns

## By Sebahattin Ziyanak, Mehmet Soyer, and Dorothy Greene Jackson

Researchers have described a wide range of effects of hydraulic fracturing on the environment and human life. Environmental factors most commonly reported are contaminated water and poor air quality. A myriad of unsubstantiated health issues has been suggested including increased cancer risks, low birth weights, increased rates of asthma, and endocrine problems. These concerns effect quality of life in affected communities. Public health nurses have a great potential for advocacy and education to help communities develop health protection measures to improve quality of life. The purpose of this study was to seek a deeper understanding of environmental and health concerns related to fracking perceived by residents of a North Texas shale gas-producing area.

### BACKGROUND OF HYDRAULIC FRACTURING

In the last decade, the natural gas industry has developed swiftly and North Texas has become a major shale gas-producing area. The Barnett Shale, located in North Texas, is one of the largest natural gas fields in the United States. However, the natural gas development in the region engenders controversies that are "filled with struggles over 'facts'" (Gullion, 2015, p.173). Hydraulic fracturing is a method of extracting oil from the ground that relies on pumping high volumes of water, sand, and solutions of chemicals into deep wells to release oil and gas trapped in rocks (Hirsch, Smalley, & Selby-Nelson, 2018; Kargbo, Wilhelm, & Campbell, 2010). Even with the highest level of precision and care by the drillers, the nature of the fracturing process impact aquatic resources and toxic chemicals exposure to the environment. Supporters of hydraulic fracturing (fracking) focus on the economic growth that gas-drilling companies are able to produce. Meanwhile, opponents focus on concerns over the health and the environmental impact from fracking because of the use of large amounts of toxic chemicals in close proximity to homes, schools, towns and cities, and restaurants (Earthworks, 2011).

## LITERATURE REVIEW

An electronic search for hydraulic fracturing studies was conducted using MEDLINE (1962–2018) and PsycInfo (1962–2018). Studies reviewed were selected based on interest of the investigators and those that explained the nature of fracturing, environmental concerns, and local appeal.

Most studies reviewed identified the fracking process as an oil and gas source resulting in water contamination and detrimental effects to the environment and human health. The proximity of the drilling to residential living areas results in public health and environmental problems (Caree, 2012). The author acknowledges the role of the grassroots claims-making activities in bringing attention to this issue. Another article concludes that: "the image of fracking as green, inevitable, and necessary is fictitious. Previous literature about communities' perceptions on natural gas development is focused on the negative environmental and health impacts of fracking (Theodori, 2009). David and Fisk (2014, p.1) found that "the opposition to fracking and support for current or increased levels of regulation are strongly related to Democratic Party identification and to pro-environmental policy attitudes."

When the documentary Gasland first appeared in 2010, YouTube became replete with videos of people living in proximity to fracking operations lighting their water on fire, ground-level ozone exposure, early-life infections, exacerbating factor in childhood asthma, recurrent details for hospital admissions among offspring, and result in premature death. People questioned the safety of these operations and their videos captured the attention of landowners and environmentalists (Hargrove, 2011). The story captured public attention and concern, and it became integrated into the community's discourse on natural gas drilling.

In parallel to the health impact, some of the main diagnosed significant health issues are asthma and various allergies. Allergies and asthma are a family matter and influences the lives of family members in numerous ways including school truancy, low academic attainment, responsive complications and postponement with the parents' day-to-day life. Isik and Isik (2017) indicate that parental education is in need of early asthma symptoms to support the children physically and psychologically for the best potential consequences. The frequency of asthma is associated with financial burden.

According to Asthma and Allergy Foundation of America (2015), asthma costs $56 billion once a year and it has been unceasingly worsening every year. Centers for Disease Control and Prevention (2015) reported that asthma is the crucial motive for truancy, and asthma-linked absenteeism generated approximately 14 million missed school days in 2013. McConnell et al. (2002) state that early-life contagions, indoor allergens, and outdoor air contaminants are more likely to cause a substantial effect on asthma occurrence.

Air quality is believed to be related to methane leaks. Known carcinogens including benzene, formaldehyde, and methylene chloride have also been reported from gas emissions from fracking (Southwest Pennsylvania Environmental Health Project, 2015). Other factors related to social and environmental health risks reported include increased intensity in diesel-truck volume traffic, swelling population growths, increased demands of local health care systems (McDermott-Levy, Katins, & Sattler, 2013).

## METHODOLOGY

Data were gathered from in-depth interviews with campaign advocates from Frack Free Denton (FFD). IRB approval for this qualitative study was granted from Texas Woman's University. We interviewed fourteen individuals for this study including a registered nurse, activists, local singer, retired professor, and local residents in Denton, Texas. This study protects participants' confidentiality and anonymity. All participants' names were replaced by pseudonyms. The data collection continued until the researchers reached the saturation point. Once the researchers generate the repetitive answers and results, we stopped engaging more participants. The interviews are semi-structured and they were held during the month of April and May 2015. The interviews were audio-recorded and each interview were lasted approximately an hour. Snowball sampling approach was performed to acquire more probable participants. Thus, interview participants were recruited through word of mouth and social media networking sites such as Facebook. Snowball sampling "begins with one or a few people or cases and spreads out on the basis of links to the initial cases" (Neuman, 2006, p. 223). At the conclusion of the interviews, we asked if the participants knew of other individuals who might be interested in taking part in the study, and if so, we asked them to forward our contact information to those individuals. Each potential participants were contacted either by phone or by email. Finally, the data were copied into Microsoft Word, organized, and printed out for analysis.

Charmaz writes "qualitative codes that segments of data apart, name them in concise terms, and propose an analytic handle to develop abstract idea for interpreting each segment of data (2010, p.45). Charmaz sees the coding is a process of describing the data. Charmaz (2010) writes, "Coding means categorizing segments of data with a short name that simultaneously summarizes and accounts for each piece of data (43)."

To be able to make a methodical explanation, coding is the gateway to generate the crucial theme of a segment. Various common themes and subjects were identified during coding data such as environmental health issues related to hydraulic fracturing. The preliminary characteristics of this study in the data are complete. After finishing the initial step for coding, obtained major themes and sub-themes were coded. For instance, the data provide environment as one of the major themes whereas water contamination, air pollution, climate change, and methane are sub-themes of it. Sample Interview Questions such as "Tell me any concerns about the safety of your community?", "Talk to me any concerns about your personal health?", "Talk to me concerns about improvement for the future?"

## FINDINGS AND RESULTS

### Public Health

The main concern over fracking within the interviewees of FFD was public health. All of the members of FFD had something to say about public health. Some provided general concerns and some provided specific evidence.

One of the volunteers of FFD, Aron, stated that "the health impacts are the same, everywhere you go. The complaints are the same and the science is starting to catch-up. The science is starting to catch-up, so that people are realizing that these impacts are real."

> Aron: So there is a lot of chemicals, a lot of resources. In just a small frack job, they use somewhere around 8000 tons of sand and that sand has to be mined and generally the frack sand mines are near some community that suffers the impacts. . . . One of the main problems are the air impacts, we're producing methane and the industry is claiming that the methane levels are falling, that's just not true. The methane levels are increasing˙ and we're four times the historical level of methane. So, the methane is what is our biggest, most imminent threat to global warming, even more so than $CO_2$. So there are some problems with it.

> Ricky: What I think the frack free people did is they capitalized on emotion as compared to fact. And they would make these health claims or safety claims or air pollution claims just like when they had the blowout. But at the end of the day, when the report comes out, no violation of air standards (Interview with Aron, April 2015).

## Asthma

Asthma is a widespread enduring illness. However, uncontrolled asthma is a noteworthy factor for truancy, frequent emergency room visits, required hospitalization, and a student's self-confidence. According to Isik and Isik (2017) these unmanageable factors can lead low school achievement and emotional complications for children as well as their parents. Asthma is one of the specific health concerns that is constructed as a health consequence of fracking by the FFD group. Some examples are provided below.

> Randy: We have an incredibly high asthma rates in Denton and I'm sure there is a relationships. So and I also know that it's very difficult, a lot of public health research is correlational, it's not cause-effect. It's very difficult to ferret out what's actually going on, but I was not mollified by the argument that, 'Well, we have a lot of asthma because we're in this metropolitan area or we have a lot of asthma because of those cement plants, south of us, the wind blowing. No, we have a lot of asthma because we have all of that plus we have the heavy incidence of the oil and gas industry (Interview with Randy, May 2015).

One of the participants, studying music, cried during the interview.

> Terry: I usually use my inhaler everyday. It's an emergency inhaler and at least once a day, I have to. I have to use it before I go on stage, before I teach music lessons just to be able to catch my breath so that I can teach my students or so that I can project my voice on stage. So yeah, it's been a challenge (Interview with Terry, April 2015).

## Allergy

Some participants from the FFD group indicated that they were suffering from allergies due to air pollution. They held fracking responsible for their symptoms. The allergy issue is another specific concern brought to light by the FFD group interviewees. Some examples are presented below.

> Kelly: No one should have to live next to something that poses this risk, but also the fact that I am allergic to air and more specifically I'm allergic to sulfur and one of the things that's released into the air with fracking is sulfur and so I can tell whenever they are fracking a well because I can't breathe (Interview with Kelly, May 2015).

> Jane: Quality of health inside of Denton for sure was my biggest issue. I have a sinus issues already and not that against totally from Fracking

because my job is pretty dusty but you know, I'm sure it contributes to it (Interview with Jane, May 2015).

## Environment

The FFD group are grounded on the potential hazards of the fracking process.

Garry: there is nothing about fracking that's good. The obvious things that bad about it are the environmental and health impacts. Its damaging to air quality in the immediate region, the process can contaminate water, it's been linked to earthquakes and all of that can negatively impact people in the direct area (Interview with Garry, April 2015).

Kerry: I am concerned about the environment. You know they always want to say that people from my side of the aisle are not concerned about that and I am very concerned about the environment but I have not seen anything that shows that fracking is going to affect the environment. No one else has been able to produce a study that shows that it affects the environment (Interview with Kerry, April 2015).

## Water Contamination

The FFD declares that water becomes contaminated due to fracking. The chemicals are released to the environment with the water that has been used to drill. Therefore, both water and the environment are affected by fracking.

Kat: They started the fracturing, all of a sudden, the water from the tank was gone because they were sucking it out for the fracking. Then two days later, the tank was full of all the water that was produced from the fracking and four cows died overnight? Horrible deaths. You could hear them mooing and crying out in the night and in the morning they were all dead . . . the veterinarians came out and said they had all died of pulmonary edema which is where all the fluid in your body goes to your lungs and they basically suffocated. So I started looking at the chemicals they were using. One of them was antifreeze, which causes pulmonary edema. So can I say that was a cause? No. Can I say as soon as the tank was filled back up with the produced water, four healthy cows died overnight? Yes (Interview with Kat, May 2015).

Phil: Some of the 19, 20-year-old kids, they just have no idea. All they hear is they want to be anti-fracking, fracking has to be banned. They think the water is bad. They don't realize we. . . . They think it's contaminating their water. They don't realize Denton gets its water from surface water, which is really outside of the. . . . Yeah, we get it from. Half comes from lake, the water supply, and half comes from Lewisville Lake. So, it's really not in the well. So, even if the wells are contaminated, some of the house in the farms, maybe they have their own well water. The Denton's water supply comes from surface water (Interview with Phil, April 2015).

Andy: You saw my kids. They don't stay in the house all day long. We have horses and stuff. We have a pond and there we go fishing. We eat the fish out of our pond. I mean, I have water well over here. They say well your water gets contaminated. I have it checked twice, and the only thing it's on high is sodium. That's I guess is common with this area. But there is no adverse chemicals because I don't know. The fracking of five wells around here. Around my place, on my place and around it (Interview with Andy, May 2015).

## Air Pollution

FFD contended that fracking results in polluting Denton's air. Therefore, another statement that supports the environmental issues is air pollution.

Kristy: It's hard to argue with wanting a healthy community, near your home especially; having diesel fumes near your home, having light on all night near your home, having your water potentially polluted and you drink that water out of the lake. So from a human health standpoint, I always emphasize that first, but then I also emphasize "this is a proximity to our property and we have no rights". These folks could come in and get around zoning and find loopholes, and find rules and Texas is big on property owner rights and mineral rights trumping surface rights is a big issue. Too many Texans, many land owners when they find that, they own 250 acres of land, that they don't own their minerals, a company could come in and set up a rig 250 feet, 300 feet from their home; I don't care what party you are, that resonates with you (Interview with Kristy, May 2015).

## Climate Change

Some participants from the FFD group indicated there is a correlation between greenhouse effects and fracking. The climate change argument is underlined by the FFD interviewees. They claim that as a result of environmental problems of fracking, climate change will become a serious problem. Here are some examples. Aron: . . . So there you have that and even the Oil and Gas industry is admitting that, climate changes really are serious so, there is that (interview with Aron, April 2015).

Charlie: They do not want hydrocarbons. I do not think hydrocarbons are a problem. They believe in man-made global warming, I do not. I mean its big, its big and its basic beliefs (Interview with Charlie, April 2015).

Kelly: I tend to view things as a socialist, as a bigger picture issue, that this is not just, Denton has this problem and Denton has to fix it, but globally. And Texas produces more than the top three other states in natural gas combined. So, we have this issue of, "we are the largest natural gas producer in the US and US is one of the largest natural gas producers globally". The 64 cubic tons produced of natural gas is from Texas, it means that we are A, consider it a sacrifice done, everyone here is considered expandable for money. And B, we are contributing on a mass scale to global warming that we can't fix. So, yeah it sucks that I can't breathe, but I've done a lot of things, for it put my health at risk (Interview with Kelly, May 2015).

Kristy: I think the pollution is definitely—they are proving that it's increasing greenhouse gases since fracking started this area; our ozone levels have gone right up. And there is a direct correlation here and other parts of the country too (Interview with Kristy, May 2015).

## Methane

The FFD group has concerns over methane gas. Methane was released during a natural gas processing. This accident was mentioned several times in the interviews as harmful to the environment.

Garry: The methane that is released by the process and the process pretty much use methane all around. . . . The chemicals and the methane and the volatile organic compounds put off by fracking can also affect people globally because it is one of the worst kinds of green house gases that we're aware of. It's much worse than CO2 so if you're concerned about the effects of global climate change, which might be described like you said as a butterfly effect where too many emissions from the fracking boom in America could lead to disastrous weather events and catastrophic climate change in other parts of the region (Interview with Garry, April 2015).

Matt: If you go 310 feet there's on below that has gas land so many wells in the early days when the water well drillers were drunk, they've make a little trip right here and kicks that, be a big flairs and they were had flairs of wells near this guy's home before any wells were ever drilled out there. The people drilling water wells accidentally going 10 feet too deep and got in to a gas land and therefore the fresh water well would burn gas and so they've been burning gas out there I mean it's so close you've got to be really careful (Interview with Matt, May 2015).

## Fracking Is (Un)safe

Drilling has been going on in the United States for over 100 years. FFD considers the fracking process as poisonous due to fracking itself.

Randy: You can't even put in a bakery 200 feet from your house, but you can bring in the oil and gas industry 200 feet from your house (Interview with Randy, May 2015).

Garry: The entire natural gas drilling process including fracking, but even before the fracking process begins, bringing in the trucks, building the apparatus, fracking and then once the fracking is done, all the collection of the materials that have to be done and even when the well is done and retired and for years and years and years afterwards, this entire process poisons life and leaves the land toxic (Interview with Garry, April 2015).

Matt: Accidents do happen. No doubt that some blowouts do happen. Do you think there are man that control people, but occasionally there will be—sometimes, you will penetrate a formation that you're going for the main formation than here the shale mostly. But all of the sudden, you find a nice, clean sandstone and it got a lot of pressure and you're drilling to it and it will kick on you. You didn't—what you do is when you start drilling, you put mud in the hole and that's to keep anything from blowing up, they have a heavy mud. The mud is heavier than the pressure down below. But a lot of time to drill down to that point, they were use real lightweight mud, not heavy weight mud. They don't start muddy enough and getting heavier weight, heavier weight until they get closer to the formation they're trying to get into. You will see that happen and occasionally, you'll penetrate this on ahead of time, and it will blow out and you got to get the people out there and get into a lake, found and then gather and they control that one well that we're talking about. But, this happens and this kind of like, because you don't want pilot to kill everybody, we understand that one. Because if it went blow out, we got a quick drill, is that make sense (Interview with Matt, May 2015).

## Being a Nurse and an Activist for Awareness

A home health nurse works as in county nursing and goes to people's homes. There were older people who lived on that street who were having more incidents of respiratory difficulty or having to use their emergency medications whenever they were fracking.

Kat: There was a big water tank out there that all the cattle were drinking out of. When they started the fracturing, all of a sudden, the water from the tank was gone because they were sucking it out for the fracking. Then two days later, the tank was full of all the water that was produced from the fracking and four cows died overnight. Horrible deaths. You could hear them mooing and crying out in the night and in the morning they were all dead. The veterinarians came out and said they had all died of pulmonary edema which is where all the fluid in your body goes to your lungs and they basically suffocated (Interview with Kat, May 2015).

Nurse would call them—what she would do is just kind of touch base with them because several of them lived alone and say to her, "It's going to be a bad ozone day and I've been to the park and it smells bad so don't come outside.

The nurse went door to door and she asked neighbors if they were experiencing any health problems. She has involved in many activities including meetings at the library, putting fliers on doors, attending every City Council meeting, having blog sites, and email groups . Additionally, they had rallies in the park.

Kat: We did everything. We talked to all our City Council and our Mayor. We even went to the County Commissioner's but they said there was nothing they could do because it was inside the city limits. And then we went to Austin to talk about the problem. Austin said, "You need to go back to Denton and talk to your elected officials. That's where you make change at a grassroots level with your local elected officials." So we came back to Denton (Interview with Kat, May 2015).

## CONCLUSION

The people who participated in the study identified concerns that were consistent with the findings in the literature. They believe hydraulic fracturing is unsafe and they believe it is making them sick with allergies and asthma. They think the drinking water and the emissions of methane may not only be affecting them but also is causing harm to the cattle. Furthermore, there is implied distrust of elected leaders as they believe the unsafe consequences of fracking will not improve.

The public health nurses in this community have an opportunity to partner with members of this community to discuss their concerns and explore what the community might want to do to improve the safety concerns.

Among the public health essentials (CDC Ten essentials of Public Health Services (https://www.cdc.gov/stltpublichealth/publichealthservices/essentialhealthservices.html) nurses help monitor the health status to identify community health problems. First the nurses could visit with concerned community members to validate their concerns. They may also invite meetings with the elected officials for assistance in describing the nature of fracking in the community: the safety levels of chemicals; required safety levels of the water; determine policy on safe fracking locations. Overall, nurses can do what is the very essence of nursing; advocate for the safety and well-being of the community by starting with "what can we do as a community to make our lives better?"

## BIBLIOGRAPHY

Asthma and Allergy Foundation of America. (2015 ). Asthma facts and figures. Retrieved from http://www.aafa.org.

Carre, N. C. (2012). Environmental justice and hydraulic fracturing: The ascendancy of grassroots populism in policy determination. *Journal of Social Change*, 4:1–13.

Centers for Disease Control and Prevention (2015). *National center for health statistics. National health interview survey.* Analysis by the American Lung Association Epidemiology and Statistics Unit using SPSS software.

Charmaz, K. (2008). Grounded theory. In J. A. Smith (Ed.), *Qualitative psychology: A practical guide to research methods* (pp. 81–110). Los Angeles, CA: Sage Publications.

Davis, C., and Fisk. M. J. (2014). Energy abundance or environmental worries? Analyzing public support for fracking in the United States. *Review of Policy Research*, 31:1–16.

Earthworks. (2011). How the Texas gas boom affects community health and safety. Retrieved from http://www.earthworksaction.org/files/publications/FLOWBACKTXOGAP/HealthReport-lowres.pdf.

Gullion, J. S. (2015). *Fracking the neighborhood: Reluctant activists and natural gas drilling.* Cambridge, MA: MIT Press.

Hargrove, B. (2011, November 24). Fear and Fracking in Southlake. *Dallas Observer News*. Retrieved from http://www.dallasobserver.com/2011 11 24/news/fear-and-fracking-in southlake.

Hirsh, J. K., Smalley, K. B., Selby-Nelson, E. M. (2018). Psychosocial impact of fracking: A review of the literature on the mental health consequences of hydraulic fracturing. *International Journal of Mental Health Addiction* 16:1–15.

Isik, E., & Isik, S. I. (2017). Students with asthma and its impacts. *NASN School Nurse*, 32 (4), 212–216.

Kargbo, D. M., Wilhelm. R. G., Campbell, D. J. (2010). Optimal well design for enhanced stimulation fluids recovery and flow-back treatment in the Marcellus Shale gas development using integrated technologies. *Environmental Science & Technology, 44(15),* 5679–5684.

McConnell, R., Berhane, K., Gilliland, F., London, J. S., Islam, T., Gauderman, W. J., Avol, E., Margolis, G. H., and Peters, M. J. (2002). Asthma in exercising children exposed to ozone: A cohort study. *The Lancet,* 359 (9304): 386–391.

McDermott-Levy, R., Kaktins, N., & Sattler (2013). Fracking, the environment, and health. *American Journal of Nursing.* 113(6): 45–51.

Neuman, W. L. (2006). *Basics of social research: quantitative and qualitative methods.* Boston: Allyn & Bacon.

Theodori, G. L. (2009). Paradoxial perceptions of problems associated with unconventional natural gas development. *Southern Rural Sociology,* 24(3), 97–117.

## Chapter Five

# Theoretical and Practical Implications

*Soyer–Ziyanak's Stages of Social Problem Model*

## By Mehmet Soyer and Sebahattin Ziyanak

The first question was "How did campaign advocates from 'Frack Free Denton' and 'Denton Tax Payers for Strong Economy' construct fracking in general?" The groups constructed their claims in the lens of human rights or property rights. The values of both groups are divergent since their interests are different. FFD's drive has been the environmental causes, whereas DTPSE's motivation is the economic advantages. Therefore, they discussed the environmental influence and whether fracking contributes to local economy in Denton. Unlike DTPSE, FFD states that fracking industry's contribution to local economy is negligible. The contending groups construct their social realities from different angels. Therefore, both groups create their claims and claims-making activities accordingly. As a result, DTPSE group interviewees did not concentrate on public health concerns. They insisted that there are no harmful effects of fracking on public health since fracking companies apply advanced technology to frack. However, some companies may not play by the game rules. That's why some explosions and other accidents occur.

In contrast to DTFSE, FFD group states fracking causes environmental harm. The claims from FFD group are grounded on the potential hazards of fracking process. However, DTPSE group claims that there is no scientific proof to demonstrate that environmental problems are because of fracking. Nonetheless, FFD asserts that water becomes contaminated because of fracking. Another claim that supports the environmental issues is air pollution. Climate change controversy is another issue indicated by the FFD participants. Each group engaged in claims-making activities to make their claims

heard in public. Both groups choose the accusatory language as their claims
as well.

One of the major claims is that DTPSE group indicates FFD activists
engage in unlawful and immoral actions, such as stealing, painting or re-
wording DTPSE's signs. Next, some members of FFD depicted as radical
left environmentalists.

> Pro-Frack Case 1: They would call in, ask for a sign, we'd go take it and put it
> in. Now we had to do several of them a lot because they kept stealing them. Oh
> we got attacked by a few of them on Facebook. One guy that kept sending me
> personal messages. To tell you the truth I didn't even reply to him. I'm not
> going to lower myself that low. If that gets you going. If that gets you off to act
> that way, then so be it. Oh, we got some dirty ass mail, but we have thrown it
> away. Where we did our mail outs and the flyer would go out?

DTPSE perceives FFD group as against for all kinds of drilling and even
for fossil fuel. There is also criticism among FFD members that FFD advisory board should embrace more diversity in their decision-making process.
Corruption is another claim that has been pointed out several times in the
interviews. DTPSE indicated that one of the reasons for passing the ban is the
presence of university students in Denton.

The second question was "How did each of these groups challenge the
claims-making activities and goals of their adversaries?" Claims-making activities are drawn from the in-depth interviews as well. FFD group recruited
volunteers from locals in Denton to accomplish their goals during the campaign. On the other hand, DTPSE worked with a private PR company in
order to raise awareness during this stage. PR Company intern hired advocates for the DTPSE cause. The claims-making activities of FFD outnumbered the DTPSE's activities. This may explain the victory of FFD in the
campaign. DTPSE has more sources, has hired PR Company, yet they lose.
This validates social construction of social problem in which preponderance
of claims-making activities foreshadows the outcome of the campaign. The
claims-makers as FFD and DTPSE participate in the claims-making activities
to construct their claims about fracking.

In awareness stage, DTPSE also made activities to promote the awareness
of pro-fracking atmosphere. In their campaign activities, they worked with
the private PR Company. All the campaign work is done by the PR Company. We sent e-mails to request an interview from the PR Company concerning the campaign details. However, they did not grant an interview. Unlike
the volunteers of FFD, the campaign workers of DTPSE are all paid. The
claims-making activities in this stage are all reflected in the respective
group's website, billboards and newspaper ads, information booth, yard sign,
TV ads, support letters form prominent locals, ads during games at Cowboy
Stadium. In policy determination stage, DTPSE also engaged in claims-mak-

ing activities to be part of the policy making stage. The claims-making activities in this stage are letters to the editor, panels, participating in city council meetings, petition booth. In reform stage, DTPSE didn't engage in any claims-making activities as a group. Instead, the oil and gas corps engaged in claims-making activities such as lobbying at Austin to pass HB40 bill. Definition of HB40 bill is that particularly prevents the regulation of oil and gas operations by municipalities and other political subdivisions.

> Anti-frack, Case 1: Our city attorney went to Austin and spoke against a bill that was going to take away- she spoke against the bill that was going to say, "If any city wants to do a fracking ban, they have to get that approved by the Attorney General first." So that means [inaudible]. She spoke on Monday and she was eloquent. She made me very proud, she said exactly "Democracy is not convenient, but you still have to allow it.

The third question was "How did the local newspaper (DRC) become the field of power-struggle of grassroots groups (Frack Free Denton and Denton Tax Payers For Strong Economy) over fracking?" Both groups have used Denton Record Chronicle as a venue to explain themselves and inform the local citizens. DRC is a powerful ground to engage in power struggle. The power struggle over fracking of both groups is located in the letters to the editor, the news coverage and advertisements in DRC.

The major claims of both groups are embedded in the letters to the editor. FFD group organized systematically to write letters to the editor for their supporters. FFD wrote more letters to the editor than DTPSE did. The claims, claims-makers, and claims-making activities of both groups are reported in the news coverage. There are also approach dissimilarities between the groups. Unlike DTPSE group, FFD group lets reporters know of their claims-making activities. The leaders of DTPSE group indicate that the reporters need to reach them, not DTPSE. Moreover, both groups paid for several ads in DRC. The major claims are also embedded in the ads. DTPSE collected ten times more political contribution than FFD made. However, the money FFD collected is from predominantly Denton residents. DTPSE mostly collected the campaign finance from oil and gas companies.

Since HB40 bill passed, the gas companies started to resume fracking in Denton. The fracking ban was overturned. This means the power struggle over fracking continues. My model recommended that the process could be cyclical and contingent upon the efforts of both groups. At this point, DTPSE left its claims-maker position to the oil and gas corps. However, FFD continues to construct the claims and claims-making activities. The supporters of FFD protest the HB40 bill in front of the drilling wells. FFD organizes demonstration to raise awareness again to show how the local control bill is disrespectful of democracy since the fifty nine percent of local voters were for the ban. Recently, the police handcuffed one of the leaders of FFD, Adam

Briggle, due to the fact that they were protesting and blocking the drilling site. During a protest, FFD activists were arrested due to the blocking trucks to enter the drilling well.

## THEORETICAL AND IMPLICATION OF THE FINDINGS

According to the social construction perspective, social problems do not come, they are made. Social problems consist of any alleged condition that comes into conflict with values or interests of a significant number of people, who agree to take some sort of action (Rubington & Weinberg, 2010).

Soyer–Ziyanak's stages of a social problem model merges the theoretical framework of value-conflict and social construction of social problems, analyzes the stages of awareness, policy determination, and reform in relation to fracking by observing the claims, claims-makers and claims-making activities in stage. However, the stages utilized in the study overlap, there are no clean-cut borderlines among the stages. For example, FFD group has been pursuing a dynamic strategy to inform and mobilize its members and local residents throughout the campaign. After House Bill 40 (HB40) passed, FFD group started to raise awareness regarding current development of fracking regulations. Therefore, potentially starting a new phase of the cycle.

The emergence of each of the three stages in natural history of social problems is dependent on the longevity of the power of claims-makers. Without sufficient and efficient claims-making, there is no assurance that a social problem will move from the first (awareness) stage to the second (policy determination) stage; or from the second to the third (reform) stage. Thus, the contending groups are the agents who can mobilize or block the necessary resources so that the process is pushed to the next stage. Table 1 summarizes the claims and counter claims of two contending groups.

According to Table 1, FFD have constructed a higher number of claims than DTPSE group. Since the stages overlap, some of these claims have been reiterated in each stage. FFD's claims are extended in various areas of concern, while DTPSE claims mainly focus on economical issues. The major claims were identified through the analyses of the in-depth interviews, which explored the claims of each group in great detail, which were categorized as claims and sub-claims. The following table illustrates the claims-making activities in each stage.

FFD's claims and claims-making activities must have been convincing for the community since 59 percent the residents voted yes for the fracking ban. In other terms, since claims-making activities are tools to persuade audiences, FFD convinced Denton residents that fracking is an actual social problem and that the fracking should be banned. Table one and two indicate

**Table 5.1. Claims vs. Counter Claims on the Effects of Fracking**

| Claims-makers | Frack Free Denton | Denton Taxpayers for Strong Economy |
|---|---|---|
| Claims and Counter Claims | Human (Citizen) Rights<br>Minimal Contribution to Economy<br>Public Health (Asthma, Allergy), Environmental problems (Water contamination, Air pollution, Climate Change, Earthquake, Methane)<br>Unsafe Technology<br>False Claims of DTPSE (Terrorism, shadowy support)<br>FFD did a good job<br>Political Corruption of the other side<br>Admitting that FFD did a good job and that DTPSE did a poor job. | Property Rights<br>Major Contribution to Economy,<br>No harmful effects on public health<br>Negligible environmental impact<br>Safe Technology<br>Claims about FFD (Eco-terrorist, tree huggers, Russian connections, fear mongering, immorality)<br>DTPSE did a poor job<br>Students are the reason to pass the ban<br>Banning all kinds of drilling<br>Admitting that DTPSE did a poor job and that FFD did a good job |

the reason why DTPSE, which spent ten times as much as FFD and sought professional help from a PR company, lost to FFD.

As a side note, DTPSE has had 10 times more funding for the activities than FFD. Moreover, both groups have pronounced FFD as the successful ally in running the campaign. One of the interviewees from DTPSE group stated that FFD's volunteers' roots are on the ground and that's why they win the battle. Also, DTPSE has been criticized for falling short in its campaign efforts although they had the resources. Moreover, their volunteers stated that DTPSE should have engaged in organizing activities for public to raise pro-drilling awareness. FFD has been successful in their claims and claims-making activities by winning the election with the help of campaign volunteers. This clarifies the victory of FFD.

## Implications

This chapter summarizes chapter1 through chapter 3. This chapter addresses the fracking debate between two groups in Denton, which became the first city banning fracking in Texas. The theoretical framework of this study will contribute to environmental and media sociology. Moreover, Soyer–Ziyanak's stages of a social problem model, merging the theoretical framework of value-conflict and social construction of social problems, will contribute to a more comprehensive understanding of value-conflict and social construction theories. This study will guide community leaders over the

**Table 5.2.   Claims-making Activities**

| Stages-Claims-makers | FFD | DTPSE |
| --- | --- | --- |
| Awareness | Canvassing | Hiring a PR Company |
| | Information Booth Websites | Billboards |
| | Blog Entries | Yard Signs |
| | Facebook Pages | Information Booths |
| | Puppet Shows | Ads at YouTube |
| | Soapbox Derby Sculpture | TV ads |
| | Light Brigade | Panel Presentation |
| | Flash Mob Dance Show | Mailings |
| | The Frackettes | Raising Fracking |
| | Documentary Film | Consciousness for Kids |
| | Screenings | Letters to the Editor |
| | Letters to the Editor | |
| | Concert | |
| | Kids in Action | |
| | Yard Signs Demonstration | |
| | Mails | |
| | Panels | |
| | Billboards | |
| Policy Determination | Attending City Council | Attending City Council |
| | Meeting | Meeting |
| | Petition Drive | Petition Drive |
| | Letter to the Editor Protest | Letters to the Editor |
| | by "Mic Check" | News Ads |
| | News Ads | |
| Reform | Calling Politicians | No activity |
| | Bus Trip to Austin | |

fracking issue that will remain ongoing for many years. Moreover, this research foster political debate between economic advantages and public health concerns in both the City Halls and Congress.

## Limitations

In this research, we met several limitations associated to generalizability, data collection process, and potential bias. Due to the nature of qualitative research, the findings of this research in Denton cannot be readily generalized to other locations. Moreover, the results may be different in other cities. Also, the fracking issue may not be generalizable across other states or countries. We collected the data at one point in time during April 2015, and our findings in this study may not detect the changes over time. In addition, sample size and a possible selection bias constitute the other limitations. The researchers' cultural and ethnic differences can be another limitation. Due to lack of trust, participants may not reveal their answers with complete accuracy during the interviews. Finally, since one newspaper was the only traditional media source in this study, it did not include other forms of media such as radio, magazines, television, or even other newspapers.

## Recommendations for Future Research

In this study, the primary data is from in-depth interviews. For future research, to do an ethnographic study of the Frack Free Denton group will explicate the activism of FFD group great in detail. Moreover, especially the website and Facebook pages of FFD is up to date. Therefore, the discourse analysis of their website and Facebook pages will explain how both group exercise their discourses. Soyer–Ziyanak's stages of a social problem model in this book can be applied by other studies.

Contest over fracking within Denton City limit is continuing saga, which must be analyzed as litigation continues within the court system. Future research is required to follow this process.

## BIBLIOGRAPHY

Rubington, E., & Weinberg, M. S. (Eds.). (2010). *The study of social problems: Seven perspectives*. New York, NY: Oxford University Press.

## Chapter Six

# International Boundaries

## *Water of the Rio Grande*

## By Dian Jordan

Shared land borders are often a stage for conflicts. The concept of shared and flowing water boundaries increases the complicated matters of dispute. This union of where land meets water is known as the riparian zone. Conflicts related to transboundary riparian areas are manifested within international power struggles related to water issues of flow control, flooding, damming, aquifers, pollution, navigation, access, and economic rights to sell. The imbalance of power related to the vital natural resource of fresh water is exhibited through multi-faceted aspects that include economies, governments, institutions, military strength, international social capital, and the geographic loci of the water sources.

This chapter will analyze the conflicts and power exhibited in a case study of United States and Mexico transboundary riparian watersheds. Interrelationships of the macro and micro structural orientations related to the shared water for these two North American countries will be examined. The comprehensive data collection contained in the Transboundary Freshwater Dispute Database (TFDD) has been utilized for this research.

Water issues are often studied as conflicts, but far less is studied on how resolutions are negotiated and maintained (Balthrop & Hossain, 2010; Diner, 2012; Dombrowsky, 2010). A number of factors influence how conflicts are framed and how resolutions are determined regarding shared international waters. Hierarchical socio-political structures and the development, application and interpretations of water laws exert a great deal of pressure on how resolutions are addressed. In some instances, no laws exist for specific issues or conflicting laws that exist. Another detriment to resolutions is conflicting economic policies between stakeholders. Early evidence of water policy in the region is explicated in the hierarchical socio-political structure of the Hohokam Indians. They acted quickly to resolve disputes around 800 A.D. among the irrigation network that served farmers for extensive miles of the Salt River Valley area located about one hundred miles north of the Mexican border. Archeological remains support the hypothesis that high level members of the villages lived on an elevated mound at key junctures of water routes. It is likely they enjoyed benefits of the water and were able to quickly identify and resolve issues amongst nearby users (Cech, 2009).

Complex and varied water valuation methodologies affect the process of resolving conflict. Issues related to these various complications are well covered in the literature, and date back to the 1800s. The United States and Mexico resolved to work together and codify boundary issues of the naturally evolving riverbanks of the basins as early as 1884 with a treaty agreement finalized in Washington D.C. for both countries. Shortly thereafter in 1889, the Convention on boundary waters: Rio Grande and Rio Colorado was established. Following five years of inactivity, the two nations began annual extensions. This process was then halted and it was not until 1944 when a more comprehensive treaty, *1944 Rivers Treaty*, was negotiated that broadened the scope of their concord to include the Colorado, Tijuana Rivers, and of the Rio Grande. In the 1960s the United States agreed to lend water to Mexico for irrigation of crops. During the 1970s and 1980s the border countries further resolved to work collaboratively on issues of salinity in the Colorado River Basin and issues of environmental pollution related to hazardous discharges. Matters of conveyance were addressed in the 1990s (Giordano & Wolf, 2002). Agreements can be made but further conflicts arise for numerous reasons. For instance, agreements related to the quantity of access from the Rio Grande were agreed upon. Quality of water was not. When it was determined pollution was occurring from the Mexican border, new resolutions had to be determined to address the pollutants. Conflicts occur when data discrepancies occur either through error or change. Significant climate change and drought conditions have sparked controversy over percentages of allocation when amounts have been over allocated in contract but are not available due to drying conditions (Dinar et al., 2007). Past treaties of resolutions are mentioned in the literature, but the details are obscure or omitted (Balthrop & Hossain 2010; Conca, 2008).

Governments and institutions support the increasingly market-driven focus prioritization over a human rights approach to water. Around the turn of the twenty-first century, Mexico vociferously complained to the United States that the increased settlement and irrigation of the western United States was affecting the Rio Grande river flow in the Juarez region. In a show of supreme political force, in 1895, the United States Attorney General ruled in favor of the United States that it held absolute territorial sovereignty over water rights. By language, this ruling precluded an absolute territorial integ-

rity positioning. This ruling followed communications between Mexican dip-
lomats imploring attention to the eroding water conditions and responses
from Washington D.C. declaring that their evidence purported dry conditions
as being the likely cause of Mexico's woes. It further reported it decidedly
was not due to massive western settlements and expansive irrigation on the
United States side of the transboundary riparian zone as the likely causes of
low water flow (McCaffrey, 1996).

Differences are underscored when international war crimes are applied as
criminal law in a traditional ex post event punitive and judicial arena, versus
applications of international laws of water that are primarily used as a tool ex
ante in the course of the negotiations. This bears out when international water
laws are a tool of political force (Eckstein, 2008). A thorough analysis of the
approval of the 1944 Rivers Treaty reveals the underlying political dimensions of the time. Former United States President Franklin D. Roosevelt had
been instrumental in bringing nations together to counter the ongoing World
War II atrocities. Pre-United Nations (UN) talks had been underway for two
years. Roosevelt was determined to build alliances in his own backyard of
the Central and North American continent nations. Acquiescing to Mexico's
demands for a water treaty would most likely ensure Mexico's support of
Roosevelt at the upcoming 1945 UN Conference to be held in nearby San
Francisco (United Nations, 2012). Negotiating transboundary water rights
improved the political atmosphere for acquiring international solidarity at the
UN level.

Conflicting economic policies and water laws between entities that share
waters can be detrimental to negotiations (Draper, 2007; Kibel & Schutz,
2007). Joint commissions and ad-hoc committees are created to address
transboundary water issues. The International Boundary and Water Commission between the United States and Mexico was designed with authority to
address flood control, hydropower, sanitation, and water storage. Newer
complications of environmental protection were addressed comprehensively
in the broader contracts of the North American Free Trade Agreement. Thusly, an additional layer of bureaucracy was created with a new institution, the
Commission for Environmental Cooperation (Conca, 2008; Frisvold & Caswell, 2000). In addition, the La Paz Agreement, the Southwest Consortium
for Environmental Research and Policy, the Good Neighbor Environmental
Board, the Boarder Environment Cooperation Commission and the North
American Development Bank have all been created within a ten-year period
from 1983 to1993 (Dinar, 2012).

In some cases of conflict, no laws exist. The underground aquifers along
the border are being depleted and contaminated by users on both sides of the
boundaries with little regard to future impacts. Draft articles have been
penned by the United Nations to codify international water law for transboundary aquifers. Resolutions ask for nebulous "equitable and reasonable

utilization and no significant harm" (Eckstein, 2011, p. 279). Each country claims domestic national laws governing their own use of aquifers. Texas, New Mexico, Arizona, and California all claim certain states' rights. Unenforceable agreements have been reached in selected locations, such as the 1999 Memorandum of Understanding between City of Juárez, Mexico Utilities and the El Paso Water Utilities Public Services Board of the City of El · Paso, Texas

Understanding the complex issues that define water conflicts requires an understanding of how water is valued. A good amount of literature is available on how water pricing and valuation is determined and the missed opportunities that result from outdated policies and practices. Upper basins for whom in the past allowed excess water to flow to lower basins now fear future demands of their own could be negated by past and current practices. A use it or lose it attitude has some upper basin regions practicing wasteful or economically irresponsible practices in order to maximize water usage, thus curtailing excess flows. These policies do not induce conservatorship or best practices for water sources (El-Ashry & Gibbons, 2009). Traditional cultural practices of the arid area were to create irrigation systems along the floodplains for agricultural purposes. Pressure is mounting to divert usage for the growing urban populations (Fernald, et al., 2007; Nitze, 2009). Eighty-five percent of global water consumption is used for agricultural practices (Jury & Vaux, 2007). As populations increase, food needs will continue to rise. The value of water for agricultural purposes cannot be understated (Goetz & Berga, 2006).

This intersectionality of influences on how water conflicts are framed and resolved sets the stage to conduct a case study analysis of the issues affecting the Mexico-United States shared waters.

A qualitative approach is utilized by conducting a single revelatory case study of selected transboundary riparian conflicts and identifying the issues of inequalities related to the Mexico-United States transboundary riparian water zones. An explicatory analysis of specific situations will be examined utilizing Ritzer's Integrative Theory of Social Analysis (Ritzer, 1991). Ritzer's theory is characterized in the macroscopic levels as manifested in the objective forms of judiciary, bureaucracy, architecture, language, and technologies of societies. The subjective influences of culture, norms and values are layers of analysis that should not be overlooked. This theory is then bilaterally influenced with micro actions of individuals that create patterns of behavior and interactions. Studying the interrelationships between macro and micro dimensions of water conflicts allows for an understanding to emerge on how the dialectical relationships shape conflict and resolution.

The Transboundary Freshwater Dispute Database (TFDD) was created and is maintained by the Oregon State University Department of Geosciences, in collaboration with the Northwest Alliance for Computational Science and Engineering. It is a compilation of full texts of 400 water-related treaties. It contains 39 United States interstate compacts in which some contain data that link to the transboundary riparian water zones along the United States and Mexico borders. The TFDD has an annotated bibliography of water conflict resolution as well the negotiating notes from fourteen case studies of water conflict resolution. The negotiating notes are particularly helpful in understanding the underlying nuances of political power and international social capital that is exerted on issues of transboundary riparian watersheds. A comprehensive news file of international water-related disputes and dispute resolutions are available and can be compared to the descriptions of indigenous and traditional methods of water dispute resolution (Wolf, 2012).

A total of 43 treaties have been recorded between the United States and Mexico that include language governing the transboundary riparian watersheds, beginning with the seminal 1848 Treaty of Guadalupe Hidalgo which ended the two-year Mexican American War and declared the international border between the countries would be the Rio Grande River.

Measuring the intensity of disputes has been coded for conflict and cooperation related to transboundary riparian water issues through the creation of a water intensity bar scale. To quantify the intensity of conflicts, the Basins at Risk (BAR) water intensity scale was created, "BAR Scale." It has unit ranges from -7 to +7 (Wolfe, 2012). The BAR Scale takes into account varying issues of conflict such as quantity, infrastructure and economic development; and levels of intensity of conflict. Declared war over water is represented at the most extreme -7 level. Declared war over water has been documented from as early as 2500 BCE in Mesopotamia over the Tigris River (Jarvis & Wolfe, 2010). Conflict at the -6 and -5 are severe negative events resulting in death and armed military involvements. Zero (0) represents a neutral stance and no significant conflicts. +7 indicates the highest level of cooperation over shared waters when states unify into one nation. The United States–Mexico situation has long enjoyed the position in the range of peaceful conflict, usually measured at the +4 to +6 levels of unified cooperation and treaties. However, serious negative conflicts have been documented between the United States and Mexico since the 1800s and continue today. Conflicts at -3 have occurred at least twice in 2001, both disputes over water quantity disbursements for the Rio Grande River. In 1989, the two countries also experienced two conflicts at -3 over water quantity involving the shared Colorado River (Wolfe, 2012).

The most recent agreement, Minute 319, considered an extension of the *1944 Water Treaty*, was signed in November 2012. The five-year agreement has been hailed as satisfactorily collaborative for both countries. Effects of drought, climate change, and expanded population growth in the arid west are the primary concerns. They have been addressed in the forms of humanitarian and environmental improvements for the Mexican parties and strengthening United States' implementations for conservation, environmental stewardship, storage, and infrastructure projects. Both countries are expecting mutual benefits from the provisions of the agreement (IBWC, 2012).

Major themes emerged in the analysis. Benefits and complications of multi-agency and multi-issue revealed the complexity of situations (Dombrowsky, 2010). Overlapping district, state, national and international water laws and traditions influenced the outcomes of conflicts (Ries, 2008; Salman, 2007). Political changes, political power, and international social capital weighted events (Blomquist, 1992; Draper, 2007). And finally, water valuation impacted negotiations (Nitze, 2009).

Explaining the positive outcomes of the multi-agency and multi-issue situation has been referred to as a diffusion of innovation (Blomquist, 1992). The methodology has elements of success that a comprehensive water plan is not capable of accomplishing with the same level of effectiveness. Blomquist specifically illuminates this microscopic analysis through the examples of working groups in the southern California region. Applying Ritzer's process of integrated theory of social analysis, the engaged parties' objective behaviors and actions included the ideas brought forth by attorneys and engineers gaining knowledge by working in overlapping regions. Board members and staffers often sat on multiple organizational structures and brought knowledge and innovation to new groups of learners. Individual water user experienced overlapping jurisdictions in respect to irrigation, salinity, or environmental protocols. Working with multiple governmental agencies created the conditions for the subjective basis defining the social construction of their realities as neighbors with shared water consumptions (Blomquist, 1992).

Conversely, there are numerous international water governance organizations with overlapping jurisdictions and purposes where the diffusion of innovation is hampered by an unseen challenge. As they attempt to mete out cooperation and compromise, they can remain unaware of silent politics that have the capability of undermining the intended work of the organizations. The strength of politics is not stagnant. Its strength is often directly correlated with the ebbs and flows with the force of economies. The stronger and more diversified the economies, the stronger the power of the country. Transboundary water organizations are often unconscious of evolving situations of power. For instance, population growth and increases in industrial usage can eventually sway political discord regarding power over quantities of water or quality of water discharges and environmental concerns (Raleigh & Urdal, 2007). Issues, mostly of access, were addressed in the earliest of agreements. Political discord occurs when newer concerns of environmental impacts and increased population consumption and industrial usage have not been as adequately addressed (Sanchez-Munguia, 2011).

In other instances, political change arrives swiftly and without notice. Transboundary water organizations, namely the International Boundary and Water Commission (IBWC), have enjoyed relative freedom from politics. The IBWC is largely staffed with engineers and field experts. Indeed, it is one of the very few United States federal institutions that are not headquartered in Washington D.C. It is based with a home administrative office in El Paso Texas. The long tenured organization, evolved from the 1944 Rivers Treaty has endured with little political interference until most recently, when in 2005 the organization's primary United States agent was appointed and then quickly dismissed by the United States President of the time. After more than one hundred years, politics finally caught up with water organizations (Mumme & Little, 2010).

According to the oft cited Harmon Doctrine of the 1890s, the ruling deduces that a country has the right to use the river water within its boundaries without any restrictions, regardless of how this use harm other countries (McCaffrey, 1996). The Harmon Doctrine was the stated position of the United States as penned by the Attorney General at the time to address the Unites States and Mexico dispute over the Rio Grande River. The doctrine generated from the power base of the upper riparian country, clearly benefited the U.S. position. However, when the United States and Canada have conflict over water, the United States becomes the lower basin country and then demands a more favorable principle of absolute. As a lower basin country, the United States desires a position that an upper basin country may not use any international river water which could have any detrimental consequences on the lower basin riparian country (McCaffrey (1996). The politics of water rights is often confounded with motives associated with the valuation of the waters.

"By treaty we had promised them [Mexico] a million and a half acre-feet of water. But we hadn't promised them usable [emphasis theirs] water" (Kibel & Schutz, 2007, p. 235). Valuation of water is not solely related to a cost per unit. The quality of water can be too high in salinity (salt) which renders the water non potable for human consumption. High salinity water can destroy agricultural fields. In 1960, the United States began draining saline water into the Colorado River, and deducted that water quantity as part of their required allocation to Mexico (Wolf, 2012). Thusly, issues of quality are but one measure in the valuation of water.

The notions of economic measures are now being computed with the ecological ramifications for wildlife and biosphere conditions. Furthermore, water valuations are more often addressing Pareto optimality. Notably, spiritual and cultural valuations, recreational valuation and associated tourism economies, and the availability of water for future generations. Usage costs must now also attempt to address and predict unforeseen expenses associated with environmental policies such as erosion control and salinity (Turner et al., 2004).

Water valuations are predicated on social issues that demand an economic analysis that balances resources and increased demands between agricultural expansion projects and urban population growth; all the while motivating conservation practices and avoidances of pollution. Expectations for traditional use must be counter-balanced with consideration for higher-value usage. The economic valuation of water includes incentive and disincentive practices. Water tariffs and pollution charges are but two examples. Furthermore, issues of cost-benefit, cost-effectiveness, and efficiencies are taken into consideration (Smuck & Schmidt, 2011).

## CONCLUSION

The United States and Mexico are two nations that share more than transboundary riparian watersheds. Discussions should be extended to beyond the limits of where the water meets the soil. Discussions should embrace the shared realities of the politics between Washington D.C. and Mexico City. They should embrace the shared impacts to migratory songbirds and other biospheric conditions. Solving transboundary dilemmas should recognize all levels of scale from the local individual that dips a handful of water from the flow to international organizations and stakeholders that construct Hoover-like dams. Contributions toward solutions should be valued by those made at the informal level. Likewise, formal agreements are crucial in defining responsibilities required from both countries (Lopez-Hoffman, et al., 2009).

Valuations of water and its usages will continually need to be reevaluated with the position of how reasonable incentives be implemented that encourage conservation, reuse, and protection of the transboundary riparian watersheds (Quealy, 2008). Furthermore, as the waters flow and change, so do the conditions in which usage and agreements are bound. Unexpected or catastrophic events such as the 2010 earthquake that imparted significant damage to the Mexican water infrastructures can cause an abrupt disruption to agreed terms and conditions. Perhaps the 2012 agreement between the United States and Mexico for their transboundary riparian watersheds will herald an era of shared waters, as well as an era of peace between neighbors and governments.

## BIBLIOGRAPHY

Balthrop, C., & F. Hossain, F. (2010). Short note: a review of state of the art on treaties in relation to management of transboundary flooding in international river basins and the global precipitation measurement mission. *Water Policy*, 12(5), 635–640.

Blomquist, W. (1992). *Dividing the waters*. ICS Press: San Francisco CA.

Cech, T. (2010). *Principles of water resources: History, development, management, and policy*, 3rd edition. Wiley: Hoboken USA.

Conca, K, (2008). The United States and international water policy. *Journal of Environment & Development* 17(3), 215–237.

Dinar, A. (2012). Climate change and international water: the role of strategic alliances in resource allocation. In *Policy and Strategic Behaviour in Water Resource Management*. Earthscan: London, England.

Dinar, A., et al. (2007). *Bridges over water: Understanding transboundary water conflict, negotiation and cooperation*. World Scientific Publishing: Singapore.

Dombrowsky, I. (2010). The role of intra-water sector issue linkage in the resolution of trans-boundary water conflicts. *Water International*, 35(2), 132–149.

Draper, S. (2007). Introduction to transboundary water sharing. *Journal of Water Resources Planning and Management*, 133(5), 377–381.

Eckstein, G., (2011). Buried treasure or buried hope? The Status of Mexico-US Transboundary Aquifers under International Law. *International Community Law Review*, 13(3), 273–290.

Eckstein, G. (2008). Examples of the political character of international water law. in the proceedings of the Annual Meeting American Society of International Law, Washington D.C.

El-Ashry, M. & Gibbons. D. (2009). *Water and arid lands of the Western United States: A world resources institute book*. Cambridge, England: Cambridge University Press,

Fernald, A., et al. (2007). Hydrologic, riparian, and agroecosystem functions of traditional acequia irrigation systems. *Journal of Sustainable Agriculture*, 30(2), 147–171.

Frisvold, G., & Caswell, M. (2000). Transboundary water management: game-theoretic lessons for projects on the U.S.-Mexico border. *Agricultural Economics*, 24, 101–111.

Giordano, M. and A. Wolf. 2002. *Atlas of International Freshwater Agreements*. United Nations Environment Programme: Nairobi Kenya.

Goetz, R. & Berga D., editors. (2006). Frontiers in Water Resource Economics Series: *Natural Resource Management and Policy*. Springer: Heidelberg, Germany.

Kibel, P. & Schutz, J. 2007. Rio Grande designs: Texans' NAFTA water claim against Mexico. *Berkeley Journal of International Law*, 25, 101–140.

McCaffrey. S. (1996). The Harmon doctrine one hundred years later: Buried, not praised. *Natural Resources Journal*, 36(3, pt2), 549–590.

Jarvis, T., & Wolf. A. (2010). Managing Water Negotiations and Conflicts in Concept and in Practice. in *Transboundary Water Management: Principles and Practice*. A. Earle, A. Jagerskog & J. Ojendal (Eds.). Earthscan: Washington DC.

Jury, W., & Vaux, H. (2007). The Emerging global water crisis: managing scarcity and conflict between water users. *Advances in Agronomy*, 95, 1–76.

Lopez-Hoffman, L. et al., (2009). *Conservation of Shared Environments: Learning from the United States and Mexico*. University of Arizona Press: Tucson AZ.

Mumme, S. & Little. D. (2010). Leadership, politics, and administrative reform at the united states section of the international boundary and water commission, United States and Mexico. *Social Science Journal*, 47(2), 252–270.

Nitze, W. (2002). Meeting the water needs of the border region: a growing challenge for the United States and Mexico. Policy *Papers on the Americas* 8, Study 1. Center for Strategic and International Studies: Washington, D.C.

Quealy, D. (2008). Bayview Irrigation District Et Al. v. United Mexican States: NAFTA, foreign investment, and international trade in water-a hard pill to swallow. *Minnesota Journal of International Law*, 17(1), 99–120.

Raleigh, C., & Urdal, H. (2007). Climate change, environmental degradation and armed conflict. *Political Geography*, 26(6), 674–694.

Ries, N. (2008). The (almost) all-American canal: Consejo de desarrollo economico De Mexicali v. United States and the pursuit of environmental justice in transboundary resource management. *Ecology Law Quarterly*, 35, 491–530.

Ritzer, G. (1991). *Frontiers of social theory: The new syntheses*. NY: Columbia University Press.

Sánchez-Munguía, V. (2011). The US–Mexico border: Conflict and co-operation in water management. *International Journal of Water Resources Development,* 27(3), 577–593.

Salman, S. (2007). The Helsinki rules, the UN watercourses convention and the Berlin Rules: Perspectives on international water law. *Water Resources Development,* 23(4), 625–640.

Smuck, G. and K. Schmidt, K. (2011). Water project toolkit. Rome, Italy: European Union Water Institute.

Turner, K. et al. (2004). Economic valuation of water in agriculture. FAO Water Reports. *United Nations Food and Agriculture Organization*: Rome, Italy.

United Nations. (2012). History of the United Nations: San Francisco conference. United Nations. New York, NY.

Wolf, T. A. (2012). Spiritual understandings of conflict and transformation and their contribution to water dialogue. *Water Policy,* 14, 73–88.

## Chapter Seven

# Supreme Justice

*The Red River Ruling*

By Dian Jordan

## INTRODUCTION

This chapter explores a case study that reveals the historical context of American Indian water rights for two tribes in Oklahoma and ultimately whether Oklahoma and the tribes have advantageous Red River water rights over another state, Texas. This case study provides an analysis to consider important areas of microscopic and macroscopic events and activities that relate to both subjective and objective examples of behaviours, actions, policy and law that identify the powers and conflicts specifically related to the Oklahoma and Texas dispute regarding water allocations from the Red River.

Not only is the Red River the boundary line between the states of Texas and Oklahoma, the Red River waters of the dispute are geographically situated within the current boundaries of the Choctaw and Chickasaw sovereign tribal nations in Oklahoma. This investigation recounts selected treaties with the tribes, the Red River Compact, selected legal rulings, and a summation of events leading up to and including the recent U.S. Supreme Court case Tarrant Regional Water District v. Herrmann et al., No. 11-889 (Supreme Court of the United States, 2013).

Beginning with the treaties enacted with American Indian tribes during the eighteenth and nineteenth centuries, we can see evidence of how issues of power and property conflicts began. This is also the beginning time of available written legal precedence of power and control related to American Indian Nations. Before the treaty era, a number of tribes claimed the southeastern portions of the United States as their homelands. Anthropologists have recorded oral histories that point to the possibility of millions of Indians living in the Americas in the pre-Columbian period, prior to 1492. The arrival of Columbus and European explorers brought diseases that decimated the Indian populations that lacked immunities or genetic tolerances for smallpox, measles, and other ailments. In part, this annihilation gave rise to the ethnocentric belief that Europeans were superior to tribes of Indians. In the 1600s, pre-Colonial European explorers brought settlers that found the new land inhospitable in regard to their ability to recognize and find edible roots, plants and berries. Early European settlers depended on the American Indians for assistance and cooperation. Soon, the Europeans, greedy for land to homestead and farm, massacred the Indians, and "forced them to relocate to reservations where land was worthless and uninhabitable" (Yang, 2000. P. 73). Relocation of American Indian tribes to Oklahoma was secured through a part of the hundreds of treaties and agreements that proliferated from 1778–1883.

This chapter takes an extensive historical analysis of how water rights for Oklahoma Indian tribes have been grossly omitted from the policies of water management leading up to the Red River Compact. The 1830 Treaty of Dancing Rabbit Creek between the Government of the United States and the Choctaw Nation ceded to the Choctaws land in southeastern Oklahoma in exchange for their lands in Mississippi. The treaty states "Wherever well founded doubts shall arise it shall be construed most favourably towards the Choctaws" (Kappler, 1904, p. 314). In the two centuries following those words, numerous subsequent treaties, documents and agreements have been in effect.

We can see the micro influences of particular individuals that affected policies toward American Indians. Vine Deloria, Jr. (1933–2005) was an American Indian scholar, prolific writer, and educator. His works exposed many twentieth century Americans to the circumstances of Indian cultures (Echo Hawk, 2010). In 1933, President Franklin D. Roosevelt appointed Harold Ickes (1874–1952) as Secretary of the Interior. Ickes' wife, Anna, spoke Navajo and wrote a book about her experiences with the tribe (Crum, 1991). Ickes, born of lower socioeconomic status, championed the rights of American Indians, African Americans and the disenfranchised. Felix Cohen (1907–1953), of Jewish ancestry, received a Ph.D. from Harvard and a law degree from Columbia. Cohen joined the Interior staff and wrote the Handbook of Federal Indian Law in 1942. Initially, the treatise was to be produced as a co-effort between the Interior department and the Justice department as an Indian Law Survey project. Cohen, as lead author of the project, was fired from his position for reasons that were never quite clear although conjecture points to anti-Semitism. "Pervasive anti-Semitism experienced by Cohen may have given him profound empathy with Indian tribes and their desire to avoid assimilation. Thus, anti-Semitism may have inadvertently helped motivate positive developments in federal Indian law" (Washburn, 2009, p. 3).

Subsequent editions of Cohen's text are still in use today. Cohen, known for his anthropologic studies and pluralistic philosophy, advocated an interpretation of the law that appreciated diverse understanding of cultural needs and values. Buttressing the realities of individuals such as these agents for social change against the macro conditions of the conflicting cultures and the powerful bureaucratic administration reveals the depth and complexity of the problems related to addressing American Indian rights.

Beginning in the 1960s and 1970s, a revival of tribal identities begins to emerge as independent tribes unite their political efforts. The pan-Indian movement is popularly described in The Return of the Native: American Indian Political Resurgence by Stephen Cornell (1998). He explicates the social deconstruction of Indianness through the centuries led to the contemporary emergence of tribes working together in order to increase their political power. Cornell agrees a sense of supratribalism, or pan-Indianism emerged when distinct tribes began to realize they needed to put aside their individual tribalism, and band together as supratribes to defend their lands and waters against white encroachment. Reiterating beliefs that Indians were sub-human beings, it allowed whites to claim rights to Indian resources without guilty consciousness, according to Cornell. These micro-subjective beliefs explain the social construction of reality as whites wished it to be. Hence, treaties and programs of United States governmental agencies effectively curtailed remaining vestiges of political powers of any tribe.

The Red River Compact is a document between the member states of Oklahoma, Texas, Arkansas, and Louisiana. Promotion of interstate comity is one of the principal purposes of the Compact. It is designed to address equitable apportionment, water quality and pollution, conservation, and flood control. It is written that the Red River Compact is intended to address water disputes in a manner that would remove controversy that might result in water rights litigation between the member states (Texas Statutes, 1979). The Compact is a result of the Water Code Title 3. River Compacts, Chapter 46. Red River Compact, Section 46.001 was signed in 1978 by the commissioner for each of the participating states. Each state's commissioner was appointed by the Governor of the State they represented. The compact was then approved by R.C. Marshall, representing the United States.

The Texas Natural Resource Conservation Commission is compelled, through language of the Water Code, to cooperate and furnish factual data related to the waters (Texas Statutes, 1979). With a principal purpose of promoting interstate comity and to remove causes of controversy, the Compact has specific provisions delineated: a) Water uses are subject to availability of water b) Any state that does not use allocated water is not deemed to have relinquished or forfeited their rights to such use c) The Compact does not allow for the interference with the rights of a signatory state to regulate water within its boundaries. Other writings have discussed the technical and commerce questions related to the water authority (Andrew, 2011; Chapman, 1985; DuMars & Curtice, 2012; Maule, 2009; Willingham, 2009).

Choctaw and Chickasaw Indian tribes have sovereign nation rights which impact disputed water allocations of the Red River Compact. Erosion of Indian power is best defined from the period of forced relocations that removed tribes from their original homelands to Indian Territory. Collectively, these forced marches are described as the Trail of Tears and occurred during the 1830s through 1840s. The relocations were conducted under harsh conditions without adequate supplies or means of transportation (wagons and horses). Significant numbers of Indians died along the route suffering from exposure, illness and disease, and lack of adequate food.

> It was then the depths of winter. . . . The Indians brought their families with them; there were among them the wounded, the sick, newborn babies, and the old men on the point of death . . . the sight will never fade from my memory. Neither sob nor complaint rose from that silent assembly (Tocqueville, [1835] 1966, p. 199).

It was not the tears of the Indians for which the forced relocations are identified, but the tears of the witnesses to the utter brutality and insensitivity, such as described by de Tocqueville. The Trail of Tears is one of the most central facets that delineate a new life, such as it was, for tribes forced to relocate to a land they were unfamiliar with. Yet, agreeing to relocation was to guarantee their sovereign status. One of the boundaries for this new homeland was the Red River. The events of relocation are deeply embedded in perceptions, beliefs and the social construction of realty for descendants. Additionally, the historical perception of American Indians and the colonial treatment of tribes by government policy and agents is evident in the current Red River water dispute.

A major weakness of the Compact is its omission of Indian rights from the apportionment of water in the Red River (Chapman, 1985, p. 88). Furthermore, the Red River Compact does expressly state "Nothing in this Compact shall be deemed to impair or affect the powers, rights, or obligations of the United States, or those claiming under its authority, in, over and to water of the Red River Basin" (Texas, 1979). This language can be inferred as the impetus that the role of the United States in the Compact negotiation is to continue its duties as protector of the tribes' best interests. "These clear terms [of the Red River Compact] protect the Treaty rights in, over and to water that the Nations have asserted" (Burrage, 2013, p. 2).

The guiding research question is why or how do water usage agreements of the Red River Compact grant favorable benefits to one state over other participating states of the Compact? Namely, does an upper basin partner, Oklahoma, have a favored benefit compared to a lower basin partner, Texas?

A review of the historical treaties, agreements, statutes, and legal proceedings reveal the depth of conflict between powerful agencies, marginalized groups, and organizations affiliated with water disputes surrounding the Red River. Access to documents is available for analysis that previously has been difficult to obtain. Treaties, compacts, legal documents, memorandums, letters and court records have typically been physically housed within the geographic location of their jurisdiction. The Internet has made access to documents available that allow for comparison and analysis of said documents that heretofore were not easily obtainable.

## RITZER'S THEORY OF INTEGRATIVE SOCIAL ANALYSIS

Ritzer's theory of integrative social analysis was applied to the macro-objective influences of law, bureaucracies and language as well as macro-subjective elements of culture, norms and values. Additionally, patterns of behaviors and actions were uniquely tied to perceptions, beliefs, and social constructions of realities. The sociological value of following Ritzer's theory for these analyses was to understand how the centuries of interactions and events came to influence the water conflict of the Red River Compact.

### Data Analysis

Data for this analysis consists of principal documents that trace the ownership and usage provisions for water of the Red River in Oklahoma and Texas. T he 1830 Treaty of Dancing Rabbit Creek is a fundamental document that ceded Oklahoma to the Choctaw Nation. It outlined the rights and benefits the Choctaws were to receive for accepting the Treaty. Treaties were initially construed under the direct auspices of the War Department (Army).

In addition to historical documents that portray the emotions and conditions of a particular time that help us to understand the differences between American Indian and white perspectives, this case study included analysis of today's issues and considered the Red River Compact, a more technical and legal document. It is directly associated with the court documents related to and including Tarrant v. Herrmann. Documents are generally prepared for use by others, and not primarily for the promotion of the case study. As the principal investigator of the case study analysis, interpretation of the contents is conducted from the approach of discovering patterns, power structures, context, inequalities of representation, omissions, and accuracy of water access for identifiable users. The process of analysis begins with detailing a timeline of documents constructed that relate to the Red River waters. Multiple sources of data are examined and categorized to better yield identifiable patterns that support or refute whether agreements grant favorable water benefits to selected users over others and the stated or implied positions of the articulating parties.

Primarily a technique of pattern matching logic determines whether predicted patterns of favoritism are demonstrated. Pattern matching allows for comparing empirically based patterns with predicted ones. The predicted patterns of variables included indicators, or lack thereof, of preferential conditions that benefited selected stakeholders. Patterns were revealed through the analysis of six different components: frequencies, magnitudes, structures, processes, causes, and consequences (Babbie, 2011). With pattern matching, an explanatory narrative is constructed to illuminate insight regarding the processes of public policy as it relates to affected populations. These patterns are explained through narrative.

While not necessarily precise, the value of narrative reflects the propositions for causal links, that in turn, are beneficial in recommendations for future policy implementations or contribute to building sociological theory. It is the shaping and ordering of the experiences that allowed us to understand the actions of the actors. In turn, organizing the events to show the connections and consequences of actions and events over time further the understanding (Chase, 2011). C. Wright Mills identified biography, history, and society as the trilogy of components for narrative inquiry (Chase, 2011). Narrative reveals the changing contexts of power and constraints of the interdependencies regarding water resource management practices of the Red River.

## Findings

Applying Ritzer's theory of integrative social analysis to the documents and actions related to the water dispute of the Red River reveals the macro-objective position of the Tarrant Regional Water District (Tarrant). This is a Texas state agency that is charged with providing water to metropolitan communities in the Dallas-Fort Worth Texas area. The specific conflicts with Oklahoma over allocation of water began in early 2007. Tarrant claimed they should be allowed to "reach" into Oklahoma to appropriate water. Oklahoma disagreed with this assertation and the Oklahoma Water Resources Board (OWRB), acting under their authority, denied the water appropriation application. Rudolf Herrmann serves as chairman of the board for OWRB, and hence, is named as the lead defendant in the case filed by Tarrant. Attempts at conciliation failed. In 2009, the United States District Court, in the Western District for Oklahoma ruled on the case in favor of Oklahoma. Upon appeal, in 2011, the case was heard in the United States Court of Appeals, Tenth Circuit. Again, Oklahoma prevailed. Still determined to have Oklahoma's water, Texas sought an audience with the United States Supreme Court. During the 2009–2010 court session, 8,159 cases were received on appeal.

They agreed to hear arguments for a mere 82 of them (Black & Boyd, 2013). Does a correlation exist for which party might receive the most benefit from having the Supreme Court hear the case? Factors exist for determining the likelihood of the case being heard. One reason is the "importance of selecting cases to resolve legal conflict" (Black & Boyd 2013, p. 1127).

Thusly, in January 2012, with avenues of reconciliation between the states appearing to be closed in the case of Tarrant v. Herrmann (2013), the U.S. Supreme Court petitioned for a writ of certiorari (sersh-oh-rare-ee), which is an order from a higher court to a lower court to send all the documents related to the case proceedings in order for the higher court to review the lower court's decision. This action was quickly followed with amicus briefs (friends of the court filings) and distributions for conference. One such brief was filed on behalf of the Chickasaw and Choctaw Nations. Although they have their own legal battles with the State of Oklahoma regarding disputes over the waters of Sardis Lake, the Nations submitted a brief on behalf of Oklahoma asserting that certainly the member state of Texas (Tarrant) had no rights to reach into their tribal nation territory for water (Burrage, 2013).

Strong evidence of outside interest in a case signals the Court that the case has importance. "Presence of interest group support is especially useful in leveling the playing field between litigants with a resource advantage (i.e., the "haves") and those that are resource poor (i.e. the "have nots") . . . weak litigants' briefs are likely to be less well argued . . . the presence of amici [friend of the court] can help make up this difference" (Black & Boyd 2013, p. 1128–1129). A total of nineteen briefs were filed in this case. In a somewhat unexpected action, the Court invited Donald B. Verrilli, Jr. Solicitor General (SG) Counsel of Record to express the views of the United States' position for the case regarding Oklahoma water rights through a brief. The justices highly value when the SG opines. The SG is often referred to as the tenth justice (Black & Boyd, 2013). Verrilli, Jr. was succinct in his analysis of the tribal position "Accordingly, water rights of the Tribes [Chickasaw and Choctaw] may be relevant to the amount of excess water available" (Verrilli Jr, 2012, p. 20). The jurisdiction and rights of the waters flowing from the Indian Territory (Oklahoma) cannot ignore tribal treaty obligations.

Previously, Tarrant attempted to buy water from the Choctaws and Chickasaws. This attempt was unsuccessful (Sotomayor, 2013). This action follows the logic if Tarrant attempted to purchase the water, then surely the water was not considered to be in the apportionment defined by the Red River Compact. When negotiations for purchase failed, the tactic was amended. Tarrant now claimed the water was in fact, theirs for the taking and pursued that interpretation for the Red River Compact. Additionally, in 2009, Tarrant entered into a memorandum of understanding to buy Red River water from the Apache Tribe of Oklahoma (Chalepah & Oliver, 2009). Tribal water rights are at the core of Winters v. United States, (1908) a Supreme Court case of the 1907–1908 session that is heralded as a doctrine that supports Indian reserved water rights. In essence, it claims the "establishment of an Indian reservation carries with it a reservation of water" (Cosens, 2012). The Court wrote that "ambiguities occurring will be resolved from the standpoint of the Indians" (Henderson, 2011, p. 576). This case study explores the question of whether, when it comes to the Red River Compact water dispute, one group is treated more justly than another in the policy and practice of water resource allocation.

## DISCUSSIONS

Ritzer's theoretical approach of integrative analysis helps us understand water resource related conflicts. It does so by providing a methodology that supports the ability to take apart the whole of the conflict from both microscopic and macroscopic aspects of history, time, actors, and events; then placing the contexts into individually integrative pieces. The pieces can be observed more precisely when not hidden within the overall context of a situation. It more fully informs us how the pieces bear influences on each other and how those influences are acted upon by individuals, agencies, organizations, and nation states. Hence, through this new understanding, we gain knowledge in understanding overt as well as underlying motivations and ramifications of the water issues, negotiation, and legal recourse that often results in unfair rewards to one party and disadvantages to another. With this greater understanding of knowledge, greater potential to reach peaceable and lasting agreements is more likely.

Conversely, as Ritzer's theoretical approach illuminates the challenges presented on water discourse, this case study can equally inform us about Ritzer's theory. Rich and thick content analysis presents an in-depth analysis for numerous microscopic and macroscopic elements of evidence for any social phenomenon. This case study presents a many-sided examination of an event from more than the perspective of current issues on the surface. Although a case study is often perceived as not generalizable, by skillfully incorporating Ritzer's theory, the methodology could be applied to numerous case study investigations that seek to understand differences in nature of behavior between groups or societies as a whole.

Furthermore, this case study establishes piece by piece how Ritzer's theory illuminates the understanding and informs social analysis. The macro-objective and macro-subjective contexts for the causes and conflicts related to water conflict of the Red River Compact. This is evidenced from our earlier review of the historical American Indian Treaties and U.S. government Indian programs. We begin to understand how Ritzer's methodology informs the situation. Each historical document lends an understanding to the macroscopic, often underlying, conditions of society, language, culture and norms that enlighten us to the hidden values of meaning evidenced in individual perceptions, beliefs, and patterns of behaviors.

We can examine the actions of individual actors and review their written records that originally were not intended for use by a particular audience, including the investigative researcher. Additionally, we can review document from the various actors and note the changing tones of description. We find that prior to removal; the Choctaws were generally regarded as having peaceable relations with encroaching nation states. Prior to the 1803 Louisiana Purchase, Pierre-Joseph De Favrot (1749–1824) was a French and Spanish soldier assigned as commandant of the Spanish military fort in Baton Rouge Louisiana, land held by the French. In the spring of 1780, De Favrot accounted for a typical supplying of trade for "ammunition for the savages" and again in the summer of 1780, he reported muskets and gunpowder were "Given to a Choctaw who brought back a horse that had been stolen belonging to the King" (De Favrot, 1780, p. 3). It is unlikely De Favrot ever imagined his accounting would be scrutinized more than two hundred years later. As it relates to the Choctaws, the accounting records what Ritzer defines as micro-subjective perceptions and beliefs. These perceptions and beliefs are revealed within the macro-objective contexts of language (savages) as well as the intersection between these conditions and the micro-objective example of patterns of behavior, action and interaction. It can be considered from the second accounting of whether Choctaws returning "stolen" horses for goods was a routine occurrence or an anomaly of the day's activities. We do know that Indian interactions with whites were becoming more frequent and the supply trading posts built along the Red River increased the frequency of these interactions.

As white encroachment intensified over the decades, relations with American Indians deteriorated. George Catlin (1796–1872) was an ethnographer of the American Indians from the period of 1832–1839. Catlin wrote from his field notes and letters how the Indians were greatly abused in their contact with the white traders. The traders sold the Indians whiskey, exacted exorbitant prices from the Indians, and in exchange offered the Indians paltry prices for the fur trade of buffalo, buckskin and beaver pelts (Catlin, 1995). This micro-objective description of a pattern in trade practice behavior infers this method of trade abuse has become quite common and has evolved into a macro-subjective form of trade culture and devaluing of Choctaw respect. This is a much harsher description of relations from the 1830s when compared to the previous trade encounters reported from the 1780s.

In addition to examples of trade and commerce, we find point by point of interest when identifying the social analysis revealed from macro-objective military actions of bureaucracy, emerging law and language. This evidence is exposed from a series of letters between the U.S. government (War Department) and the Choctaws. The letters are written over the summer of 1830, prior to the September signing of the Treaty of Dancing Rabbit Creek. Eaton, serving as U.S. Secretary of War, and considered a primary Indian Agent for the Choctaw tribe, wrote to the Governor of Georgia on June 1, 1830. At the time, Georgia was urgently pressing for Choctaw removals from Georgia in order for white settlements to proceed without conflict with Indians. In the letter, Eaton asks in regards removing the Choctaws to Oklahoma "Is this injustice and cruelty? Assuredly, it deserves a milder name" (Eaton, 1835, p. 2). And on the very same day of June 1, 1830, Eaton pens a letter to the Choctaws wherein Eaton describes the removal in far different terms. "Go beyond the Mississippi, where you can be under your own laws, and upon your own land, with none to interrupt you. . . . Send word to your neighbors, the Chickasaws, that they may make a treaty and remove with you" (Eaton, 1835, p. 4). A few weeks later, Eaton is still writing letters. This time he writes to military comrades (Indian agents) in anticipation of the signing of the 1830 Treaty of Dancing Rabbit Creek. He directs the agents to carry the letter to the Choctaws. "Every Indian . . . must perceive that they cannot live happily within the States . . . subject to laws other than their own. . . . The President would gladly avert such a state of things, and see his red children placed in a situation where they could enjoy repose and be happy" (Eaton, 1835, p. 75).

## SOCIOLOGICAL IMPLICATIONS

The sociological implications of analyzing documents reveal patterns of power and inequality from a macro-objective and micro-subjective frames (Ritzer 1991; Yin 2014). For example, the language and writing in the Treaty of Dancing Rabbit reveals it was signed by the United States government, represented by Secretary of War, John H. Eaton (1790–1856) who served in his position as Secretary from 1829–1831. Also representing the United States was Colonel John Coffee (1772–1833) a troop commander. One hundred and seventy-one "chiefs, captains, and head men of the Choctaw Nation" signed the treaty on their tribe's behalf (Kappler, 1904). Of those 171 men, eleven were capable of signing their name in English. The remaining 160 Choctaws signed an "X" indicating they understood and agreed to the terms of giving up their homelands in Mississippi in exchange for lands in Indian Territory (Oklahoma). The micro-subjective social construction of reality for the tribal representatives was their perceptions of their positionality for negotiation. This perception was influenced by the events leading up to the Treaty. As westward expansion was encroaching from the eastern seaboards into the State of Mississippi, the Choctaw homeland, the Choctaws were experiencing micro-objective patterns of behaviors and interactions from the new settlers.

> Indians collaborated with Europeans in economic and spiritual exchanges . . . it is not consistent with the dominant view of European motives. . . . We are still faced with the image that they possessed such a concentration of technological superiority and material wealth that they had to expand outward and conquer tribal peoples (Miller, 1993, p. 346).

These negative interactions were compounded by macro-objective and macro-subjective examples of campaigns by the government and mass media to cudgel the Choctaws to submit to vacating their fertile Mississippi lands and to accept offers of the new land, Indian Territory. President Andrew Jackson, who served the presidency from 1829–1837, spoke to the U.S. Senate espousing that forced removal of the Indians would incalculably strengthen the southwestern frontier and enable those states [Mississippi] to advance rapidly in population, wealth, and power (Jackson, 1830). Far stronger than innuendo, Jackson iterates the government's position is to save the Indians and relocation is their only recourse to prevent utter annihilation. Hence, The General Government kindly offers him a new home and proposes to pay the whole expense of his removal and settlement (Jackson, 1830).

Conversely, as Jackson and others harbored imperialistic attitudes toward American Indian tribes, there were individuals that recognized the systematic treatment of disadvantage enacted upon the tribes. As stated, Problems of Indian Administration , by Miriam, was a document that indicated American Indians had been dispossessed of their lands. The seminal work resulted in, among other broad policy changes, the termination of programs related to how American Indian lands were allocated as part of the 1934 Indian Reorganization Act (Singletary & Emm, 2011).

Sociological implications of the way in which the American Indian tribes were forced to reconsider their traditional ways in order to survive within the new restrictions placed upon them are momentous. American Indians were forced to occupy lands that had somewhat different flora and fauna from what they were accustomed to in their native homelands. The Anglo expansion west affected the ecosystem in profound ways "Emigrant parties, particularly their herds, consumed the grasses, seeds, and game that sustained . . . Indians. Water sources also became jeopardized, not so much by outsiders' consumption as by their animals' defecation" (Blackhawk, 2006, p. 249). These events significantly impacted American Indian tribal ability to maintain normative macro-subjective elements of their culture (bison hunting) as well as micro-objective patterns of behavior for how they obtained clean drinking water. On another level, an interaction pattern of macro subjective and micro subjective evidence is portrayed in the recounting of the oral history interview of the elderly Mary McDaniel, of Hunkpapa and Oglala descent.

> When I was a little girl at Cheyenne River. . . . We used to have to go down to the creek every morning. My grandmother talked to the water. Before we washed, she would tell it how beautiful it was and thank it for cleansing her. She used to tell me to listen to the water, and then she would sing a song to it (Shorris, 1971, p. 192).

The story underlies the importance of claiming a qualitative sociological approach to the analysis which is afforded through case study. "Data tends to sanitize; real people become abstractions, and the true richness of water and its connection to life gets lost. . . . It is not possible to understand the settlement era [1978–current] without looking at the antecedent conditions and events" (McCool, 2002, p. xii; McCool, 1994). Her story is not sanitized when told in her own words that demonstrates how Indian culture, norms, and values regarding water are passed down to the next generation through patterns of behavior, action, and interaction.

Another example explaining the decline and rise of American Indian circumstances involves education of the youth. Oral history was not a favored educational tool of the American education system. During the late 1800s and early to mid-1900s, Indian children were often educated at boarding schools, funded and operated by the U.S. government. The education programs were examples of a macro-objective configuration relying on bureaucracy, language, and architecture promulgating acculturation to white society. As Indians became educated, these young leaders began to form national political organizations, such as the Society of Native Americans, founded around 1911. New generations of Indians began to see survival as dependent on their ability to reach self-sufficiency within the white context of society.

Although not always successful, it has long been the United States federal government's objective to care for the American Indians in order to assist their expected assimilation into white culture. In this regard, the current case of Tarrant v. Herrmann (2013) was not the first time Oklahoma and Texas battled over the Red River. The following case highlights an example of the United States taking a positive position for Indian rights. The case demonstrates macro objective and micro objective integrations. In the late 1800s, prior to Oklahoma gaining statehood, the United States sued Texas in order to absolutely and finally resolve a dispute of land ownership that, in essence, arose from the erroneous mapping of the boundary river. The conflict arose from errors and omissions that are resultant from unintended consequences of actions by individuals. In 1852, under the leadership of Captain R.B. Marcy and Lieutenant George B. McClellan, one hundred and twenty soldiers explored the headwaters to map the Red River. They mistakenly traversed the north fork, rather than the south fork, the south fork being the main channel. The disputed riparian land between the forks of the Red River included about two thousand acres of valuable land. White citizens, believing they were homesteading in Texas, began to settle the land between the forks. The dispute led to the United States Supreme Court case, United States [in representation for the Indian Territory land] v. Texas, 162 U.S. 1 (1896), for the argument of whether the disputed Red River riparian lands belonged to Texas or should remain reserved as Indian land. The Court sided with the United States and awarded what is now Jackson, Harmon and Greer counties in Oklahoma to the Indian Territory (Morrison, 1987). In this analysis, improved technology and methods for surveying revealed the mapping errors and it prompted closer scrutiny of the boundary laws. Yet, through mistaken beliefs about land positionality, a pattern of behavior and actions led to encroachment onto Indian lands. Analyzing the data through Ritzer's integrative theories illuminates the inescapable connections between infrastructures and agencies.

When discussing the event of Choctaw removal, Eaton questions it as "assuredly" not unjust and cruel when speaking to the President of the United States. And in almost the same breath, describes removal as an event of happy and joyful repose in his communication to the Choctaws. Furthermore, Eaton goes on to describe the conditions of which the Choctaws will receive the lands in Oklahoma.

> Brothers, listen . . . [Oklahoma land is] in all respects equal, if not superior, to the one you have. Your great father [President of the U.S.] will give it to you forever, that it may belong to you and your children . . . free from all interruption. Brothers; there is no unkindness in the offers made to you; no intention or wish is had to force you from your lands. . . . The attachment you feel for the soil which encompasses the bones of your ancestors is well known; our forefathers had the same feeling" (Eaton, 1835, p. 245–246).

Eaton's language exudes the paternalistic and colonial attitude that emerges for relations with the American Indians. He repeatedly calls them brothers and refers to the President as their benevolent father who only wishes the best for their circumstances.

In response to the letter received on behalf of the President, the Choctaws respond in kind. Their letter informs us from an integrative analysis between their micro-subjective form of perceptions and beliefs as to the true intent of aforementioned correspondence and the macro-objective bureaucracy and architecture of the government. Additionally, we can frame the interactions of the parties involved from the micro-objective patterns of interaction. The Choctaw response begins

> Friends and Brothers: our father, the President, has communicated to us . . . his earnest desire to make us prosperous and a happy people . . . he proposes to give us a country . . . in fee simple, or to use his own words, 'as long as the grass grows, and the water runs' Father: your red children view this state of things with feelings of deepest regret. . . . Father: you call us your children . . . we know you are sincere. . . . But we humbly beg . . . that we have now arrived at the age of maturity . . . justice to ourselves, compel us to say to you, that we cannot consent to exchange the country where we now live, for one that we have never seen (Eaton, 1835, p. 243–244).

Clearly, the Choctaws, from a micro-subjective point of view perceive the social construction of their reality in far different terms than as the misguided red children the President prefers. Undeterred by the Choctaw response, Eaton is now joined by his compatriot Indian Agent, Coffee; they both sign their names to the response and write back to the Choctaws on behalf of their great father, the President.

> Brothers: we wish to give you a pleasant country, of good soil, good water and climate, and in extent sufficient for all your wants; and when you are gone, for the wants of your children. . . . We are advising our red brothers for their own prosperity's sake to remove" (Eaton, 1835, p. 245).

At this juncture, we are now able to distinguish how social forces from the past shaped the present conditions for water conflict. This information allows for a more accurate framing of the particulars surrounding the Red River Compact. Most notably, the principle parties are the bureaucratic agencies (Tarrant and OWRB) associated with the individual states Texas and Oklahoma, respectively. As history repeats itself, the fair interests of the American Indians have not been fully considered. It was other complex issues that clouded the Red River Compact's ability to mediate disputes of allocations and rights for the Red River Basin and its tributaries. For instance, rights to commerce, water quality, wildlife management issues, and pollution concerns that increased disputes related to water management within the Compact. However, the historical governing documents reveal jurisdiction and rights of the said waters were initially granted to the Choctaw tribe of American Indians by the U.S. government in the 1830 Treaty of Dancing Rabbit Creek.

Ritzer's theory highlights how the events of long ago at Dancing Rabbit Creek came to be embedded in the current Red River conflict. The U.S. Supreme Court justices stated one of their judicial norms is to consider the friend of the court briefs as voices for the have nots. Marginalized groups are notably the minority voice. As previously noted, the nineteen "friends of the court" briefs that were submitted to the justices for review included the following representations. Seventeen bureaucracies categorized as governmental agencies, organizations and associations; states and cities; and universities. Primarily, the briefs represented the voices of dominant society and bureaucracies (Scotus, 2013). These voices represented how the macro-objective elements of society influence conflict resolution.

## RITZER'S MACRO-SUBJECTIVE ELEMENT (JUDICIAL NORMS) ARE MICRO-SUBJECTIVE PERCEPTIONS

Intersecting with Ritzer's macro-subjective element (judicial norms) are micro-subjective perceptions and beliefs for two groups. The intersection is revealed with the ability of these groups to have an interaction with the highest court (micro-objective). It had a positive impact regarding the outcome of the Red River water conflict. While the majority of the friend of the court briefs did represent the majority voices, one brief represented a grassroots organization of primarily individual citizens on behalf of Oklahoma's position that the waters of the Red River should be retained by Oklahomans. It illustrated that water conflict resolution must go beyond the considerations of buyer and seller. The resolution would have social considerations that should not be ignored. "Citizens and communities in Southeast Oklahoma depend upon water to support the growing tourism and recreation industries" (Derryberry & Aamodt, 2013, p. 4). Tourism is directly related to the non-consumptive use of the water, such as boating and fishing. Furthermore, the grassroots brief highlighted the fact that every county in Oklahoma is currently classified as experiencing drought conditions according to the U.S. Drought Monitor Report (Aamodt & Derryberry, 2013). Although the citizens of Oklahoma were not direct parties to the Red River Compact dispute, the residents along the Red River Basin had strong opinions regarding how taking the water would affect their communities. For one community alone, 52 percent of all families with children under the age of five years live below the poverty level. For female headed households, the number increases to seventy-six percent (U.S. Census, 2010). The process of the amicus brief allowed the court to hear their voices.

The second minority voice brief was submitted by the Chickasaw and Choctaw Nations of Oklahoma. It enumerated their rights as it related to the Red River water dispute. Although tribal rights are not central elements of the case, this omission of Indian rights was identified as important by Chapman as early as 1985. Additionally, Solicitor General Verrilli, Jr. cautioned those rights should not be ignored in his response to the Supreme Court justices in 2012 (Chapman, 1985; Verrilli Jr, 2013). These examples characterize the points of macro-objective bureaucracy and micro-objective patterns of behavior associated with the reasons for conflict that built up over time. With a stated objective of the Red River Compact to address disputes that would lessen litigious activity, the Compact failed in this regard. In 2013, the United States Supreme Court addressed the water conflicts in Tarrant Regional Water District v. Herrmann (2013), having risen from the Tenth Circuit Court of Appeals.

The Red River Compact attempts to address the rights and needs of various constituents yet does not adequately address the needs or rights of tribes. One of the limitations of this case study is that it does not fully investigate all party rights to the Red River waters. It focuses solely on whether the Compact grants favorability to the upper basin party (Oklahoma), and in turn, how the erosion of American Indian tribal power affected their water rights. Furthermore, this case study did not fully analyze associated parameters of water conflict. As was portrayed in the United States-Mexico conflict, we can understand the value of ensuring that all avenues for water conflict resolution are considered and their related abilities to help reach resolutions satisfactory to all parties in the dispute. The Red River Compact disagreements over apportionment were addressed. Transboundary issues, salinity, environmental degradation, wildlife habitats, and rights to navigation are water conflict issues that were not studied. Each of these areas are ripe for further study and would enhance a fuller understanding of how shared waters must be managed from a holistic perspective.

## CONCLUSION

What this case study does highlight are the issues related to tribal water rights. Oklahoma, as does many other Western states, has unique laws governing Indians. "Congress has passed many special laws for Oklahoma tribes, especially for the Five Civilized Tribes" (Cohen, 1945, p. 425). The State of Oklahoma in efforts to regulate its water resources, has failed to adequately recognize sovereign nation rights of these tribes. These rights have generally been ignored and tribal efforts at dialogue have been rebuffed.

The Chickasaw Nation refers to the California and New Mexico states as positive examples for implementing tribal-state dialogues related to water planning (Greetham, 2008). The general attitude of western states' utter dismissal for tribal rights was popularized in Paul E. Lawson's work in the 1990's. His review resulted in the publication of When States' Attorneys General Write Books on Native American Law: A Case Study of Spaeth's American Indian Law Deskbook. The Spaeth book was, as stated by Lawson (1995):

> A self-serving collaborative effort by elected political officials to foster a symbolic gesture of concern and compassion for Native Americans. . . . The conference simply produced a document that speaks in the defense of the

American legal system . . . and justification to the continued repression of Native Americans by the American legal system (p. 229).

Assistant Secretary of Indian Affairs, Kevin K. Washburn has also lent his expertise to the annals of Indian Law. He is co-author and editor of Cohen's Handbook of Federal Indian Law, 12th edition.

In addition to tribal considerations, the ultimate analysis by the Supreme Court resulted in a final opinion on the Texas and Oklahoma case that was unanimous. With all nine justices in agreement, the Honorable Justice Sotomayor delivered the opinion which stated in part "Adopting Tarrant's reading would necessarily entail assuming that Oklahoma and three other States silently surrendered substantial control over their waters when they agreed to the Compact (Sotomayor, 2013 p. 3).

This chapter illuminates the historical context of treaties and documents that reveal how and when American Indians were dispossessed of their lands and broad policies exhibited legal subjugation over American Indian tribes. The case study further reveals early twentieth century efforts by selected actors and agencies to revert trends of patriarchal control. Following these trends, efforts at pan-Indian movements gained in popularity, revitalizing Indian pride, new cultures and power. The effect of education for American Indians was also a contributing factor to their ability to assert their position pertaining to Red River water rights through their contribution of knowledge in the lawsuit's amicus brief. Through analysis, this case study does not reveal that in review of the Red River dispute, historical American Indian water rights have been adequately addressed. For additional reading on the subject of tribal water litigation for waters in southeastern Oklahoma, follow the litigation filed by the tribes *Chickasaw Nation v. Fallin*, No. 5-11-cv-00927-W (2011) and the return volley filing Oklahoma Water Resources Board v. United States, No. 5:12-cv-00275-w (2012).

This case study explains the patterns of favorability generated through shifting powers and evolving constraints related to the interdependencies of water rights and water needs for the specified tribes, organizations, governmental agencies, and citizens residing on the northern and southern lands divided by the Red River. A macro and micro functional analysis through pattern matching exposes instances when favourable water rights are benefiting one group over another. Additionally, through a sociological approach, this case study demonstrates how Indian culture, norms, and values regarding water are passed down to the next generation through patterns of behavior, action, and interaction. And likewise, how European culture and values affected the policies and colonialism directed toward American Indians that have led us to the positions we see evidenced through water conflicts exhibited in the United States. The case study is an appropriate tool to supply the sociologist with that supports the investigator for the central task in making the connections between the micro world of events in everyday life and wider social structures and long-term processes of change (Inglis, 2010). Coupling the case study with Ritzer's integrative analysis, demonstrates the individualized elements of structure or practices cannot be wholly understood outside the practice of integrative analysis that considers the context of the larger forces when contemplating the issues of whether policies or practices favor one group over another.

## BIBLIOGRAPHY

Aamodt, B. J. (2012). Riparian rights in Oklahoma: Common law water rights. *Oklahoma Water Issues*, 2(1), 3.

Andrew, N. (2011). Interstate water transfers and the Red River shootout. *Texas Environmental Law Journal*, 41(2), 181–203.

Babbie, E. (2011). *The basics of social research*. Belmont CA: Wadsworth.

Blackhawk, N. (2006). *Violence over the land*. Cambridge MA: Harvard University Press.

Black, C. R., & Boyd, L. C. (2013). Selecting the select few: The discuss list and the U.S. Supreme Court's agenda-setting process. *Social Science Quarterly*, 94(4), 1124–1144.

Burrage, M. (2013). *Brief for the Chickasaw and Choctaw Nations as Amici Curiae in support of respondents*. Retrieved from http://www.americanbar.org/content/dam/aba/publications/supreme_court_preview/briefs-v2/11-889_resp_amcu_c-c-nations.authcheckdam.pdf .

Catlin, G. (1995). *Letters and notes on the North American Indians two volumes in one*. North Dighton, MA: JG Press.

Chalepah, E. A., & Oliver, J. (2009). Memorandum of understanding. Anadarko OK: Apache Tribe of Oklahoma. Retrieved from http://s3.amazonaws.com/content.newsok.com/documents/Exhibit2ApacheMOU.pdf .

Chapman, A. M. (1985). Where east meets west in water law: The formulation of an interstate compact to address the diverse problems of the Red River Basin. *Oklahoma Law Review*, 38(1), 1–112.

Chase, E. S. (2011). Narrative inquiry: still a field in the making. In Social and educational research. Pp. 42–59 In N. K. Denzin & S. L. Yvonna (Ed.), *the sage handbook of qualitative analysis*. Lincoln. Los Angeles: Sage.

Cohen, S. F. (1945). *Handbook of Federal Indian Law*. Washington, DC: United States Government Printing Office.

Cornell, S. (1998). *The return of the native: American Indian political resurgence*. New York, NY: Oxford University Press.

Cosens, B., & Royster, V. J. (2012). *The Future of Indian and federal reserved water rights: The winters centennial*. Albuquerque NM: University of New Mexico Press.

Crum, J. S. (1991). Harold L. Ickes and his idea of a chair in American Indian History. *The History Teacher*, 25(1), 9–34.

De Favrot, P. J. (1780). Diary of events at the fort having to do with special meals and gunpowder usage. *Louisiana Research Collection*. New Orleans LA: Howard-Tilton Memorial Library. Retrieved from http://cdm16313.contentdm.oclc.org/cdm/compoundobject/collection/LPC/id/286/rec/1.

Derryberry, L., & Aamodt, J. (2013). Brief for Oklahomans for Responsible Water Policy as Amicus Curiae in Support of the Respondents. American Bar Association. Retrieved from http://www.americanbar.org/content/dam/aba/publications/supreme_court_preview/briefs-v2/11-889_amicus_okla_responsible_water.authcheckdam.pdf.

DuMars, T. C., & Curtice, S. (2012). Interstate compacts establishing state entitlements to water: An essential part of the water planning process. *Oklahoma Law Review*, 64(4), 515–538.

Eaton, H. J. (1835). *Correspondence on the subject of the emigration of Indians*. Google e-book. Washington: Duff Green.

Greetham, S. (2008). A general proposal for tribal-state water dialogue. Email communication to Duane Smith, Executive Director of Oklahoma Water Resources Board. Ada OK: Chickasaw Nation Division of Commerce.

Henderson, T. (2011). Five tribes' water rights: Examining the Aamodt adjudications' Mechem doctrine to predict tribal water rights litigation outcomes in Oklahoma. *American Indian Law Review, 36* (1), 125–160.

Inglis, T. (2010). Sociological forensics: Illuminating the whole from the particular. *Sociology, 44*(3), 507–522.

Kappler, J. C. (1904). *Vol. II, Treaties Indian Affairs: Laws and Treaties.* Washington: Government Printing Office. Retrieved from http://digital.library.okstate.edu/kappler/Vol2/treaties/cho0310.htm.

Maule, K. (2009). When silence speaks 1,000 words: negative commerce clause restrictions on water regulations and the case of Tarrant regional water district v. Herrmann. *Texas Environmental Law Journal, 38*(4), 242–268.

McCool, D. (2002). *Native Waters: Contemporary Indian Settlements and the Second Treaty Era.* Tucson AZ: University of Arizona Press.

Morrison, D. J. (1987). *The social history of the Choctaw nation: 1865–1907.* Durant, OK: Creative Informatics, Inc.

Sotomayor, J. (2013, Apr 23). Tarrant regional water dist. V. Herrmann Opinion Syllabus 569 U. Retrieved from https://supreme.justia.com/cases/federal/us/569/614/.

Supreme Court of the United States. (2013). Retrieved from http://www.supremecourt.gov/opinions/12pdf/11-889_5ie6.pdf .

Shorris, E. (1971). *The death of the Great Spirit: An elegy for the American Indian.* New York, NY: Simon and Schuster.

Singletary, L., & Emm, S. (2011). Working effectively with American Indian populations: A brief overview of federal Indian policy. *Fact Sheet-*11-34. Reno NV: University of Nevada Cooperative Extension.

Texas Statutes. (1979). Water Code. Title 3. River Compacts. Chapter 46. Red River Compact. Retrieved from http://www.statutes.legis.state.tx.us/Docs/WA/htm/WA.46.htm .

Tocqueville, A. (1835) (1966). *Democracy in America.* J.P. Mayer & M. Lerner. New York, NY: Harper and Row.

U. S. Census. (2010, Apr 15). Community facts. Retrieved from www.factfinder2.census.gov/faces/tableservices/jsf/pages/productview.xhtml?pid=ACS_11_5YR_DP03.

Verrilli, Jr. B. D. (2012). *Brief for the United States as Amicus Curiae.* Washington: Supreme Court of the United States, Department of Justice No. 11-889: 1–22.

Washburn, K. (2009). Felix Cohen, Anti-Semitism and American Indian Law. *Arizona Legal Studies Discussion Paper No. 09-03.* Retrieved from http://ssrn.com/abstract=1338795 .

Willingham, A. M. (2009).The Oklahoma water sale moratorium: How fear and misunderstanding led to an unconstitutional law. *University of Denver Water Law Review,* 12, 357–376.

Yang, Q. P. (2000). *Ethnic studies: issues and approaches.* Albany, NY: State University of New York Press.

*Chapter Eight*

# Political Cooperation for American Indian Water Rights to Sardis Lake

## By Dian Jordan

### INTRODUCTION

Water issues are often studied as conflicts; less on how resolutions are negotiated and maintained. A number of factors influence how conflicts are
framed and how resolutions are determined regarding shared waters. I explore the power and politics regarding water practices and policy development. Understanding who makes decisions and how those decisions are made
for water rights is critical to realizing the consequences of market-based
decisions, lawsuits, and negotiated settlements. Decisions often ignore ecological and social sustainability stewardship needs. Ritzer's theory of integrative social analysis is used to present a case study of Sardis Lake. The
case identifies power and conflict between governmental institutions and
Oklahoma Indian tribes. The case builds a linking research agenda to explain
how macro and micro functions influence water discourse. In essence, sociology of water can be understood as "a practice in which structure and agency
'meet' to reproduce and transform society" (Mollinga, 2008, p. 7).

   For many federally recognized tribes within the United States, issues
surrounding the authority to sell water reside within the framework of federal, state, and sovereign nation entities. In Oklahoma, Sardis Lake is a reservoir built by the United States Corps of Army Engineers (Corps) by damming a tributary of the Kiamichi River that flows into the Red River. Geographically, Sardis Lake is situated within tribal land boundaries in southeastern Oklahoma. Chickasaw Nation and Choctaw Nation of Oklahoma v.
Fallin et. al. avows the plaintiffs' treaty rights. Named as defendants are
Oklahoma Governor, Mary Fallin; Oklahoma Water Resources Board

(OWRB); City of Oklahoma City; and the Oklahoma City Water Utility Trust (OCWUT).

The case asserts the State of Oklahoma negotiated to sell Sardis Lake, water that is arguably not theirs to sell. Tribes claim sovereign nation rights as well as proprietary interests in the waters as vested to them by federal law (Canby, Jr., 1988; Greetham, 2012; Henderson, 2011; Miller, 2012). Content and narrative analysis was conducted utilizing Ritzer's theoretical framework of integrative social analysis. I assess how treaties with the tribes, oral histories, letters, field notes, legal rulings, and summation of events leading up to and including *Chickasaw v. Fallin* led to the water dispute for Sardis Lake. Data for this analysis consisted of documents that traced interactions relating to land and water in southeastern Oklahoma. Documents include correspondence of State and tribal letters to Oklahoma governors. In addition to correspondence, the federal lawsuit, Chickasaw v. Fallin, is included as a data source; it outlines the primary claims to Sardis Lake water rights. Implications of the historic Winters decision is also considered. This 1908 U.S. Supreme Court ruling is a cornerstone for legal precedence upholding Indian water rights (Cosens & Royster, 2012; McCool, 2002).

Descriptive and chronological frameworks illuminate the "terms of multiplicity of decisions, by multiple officials, that had to occur in order for implementation to occur" (Yin, 2014, p. 140). The presentation exposes salient patterns of abuse and embedded beliefs that influence prevailing social constructions of realities. Data revealed what constituted, and who held, particular powers of authority and power structures. Additional data informed the impoundment history and the political actions that created the Sardis Lake reservoir and the subsequent lawsuits regarding rights to sell the water.

## STATEMENT OF THE PROBLEM

Tribal rights eroded as European and United States' influences dominated tribal nations and when tribal populations diminished (often through epidemics of illness and disease). These events eroded tribal powers; they were not completely extinguished. "What remains is nevertheless protected and maintained by the federally recognized tribes against further encroachment by other sovereigns, such as the states. Tribal sovereignty ensures that any decisions about the tribes with regard to their property and citizens are made with their participation and consent" (U.S. Department of Interior, 2013). The United State Constitution empowers Congress to ratify treaties and "regulate Commerce with foreign Nations . . . and with the Indian Tribes" (Article One, Section eight). These provisions demonstrate that tribes are to be considered equivalent to any other government-to-government activity. Because of the unique powers of Congress in relation to Indian affairs at the national level, tribes are relieved from a position of subordination to individual states rights.

Advocates for the Five Civilized Nations argue these tribes have even stronger reserved rights with significant treaty and jurisdiction over the water (Robertson, 2011). In exchange for their civilized negotiations with the United States, they received their lands in fee simple. Treaties expressly exempted tribal nations from state jurisdiction. Further strengthening Oklahoma Indian rights was that Congress allowed Oklahoma into statehood under the terms of the 1906 Oklahoma Enabling Act. "Oklahoma's very formation was conditioned on its agreement not to disturb tribal rights or interfere with superior federal authority. (Burrage, Rabon & Greetham, 2012, p. 17).

It was the actions of principal entities in the negotiations and proposed sale of Sardis Lake water by the defendants that disregarded these rights. Principals negotiated the sale of water that is arguably not theirs to sell. In order to answer the question of power and conflict related to negotiated water sales, I delineate how the actions and emotions of individual and group experiences came to be embedded in the discourse related to Sardis Lake.

## CASE HISTORY

In 1962, the Flood Control Act granted authorization for the U.S. Army Corps of Engineers to construct Sardis Lake. The purpose of the reservoir was for flood control, water supply, recreation, and fish and wildlife. Although authorized, construction did not begin in earnest until 1974 when the State of Oklahoma, contracted with the Corps, to make regular annual payments for thirty years to pay for the project. There were no suggested payments to the tribes. The Indians watched as construction began through their territory, reminiscent of previous experiences. Fulfilling part of the design function, Sardis became a recreational destination for crappie fishing and trophy bass with over one hundred miles of shoreline and covering nearly fourteen thousand acres (USACE 2019).

The metropolitan area of Dallas-Fort Worth Texas recognizes the need to provide critical water for a growing urban population (OWRB 2010). The metropolis lies about two hundred miles south of the Red River. Oklahoma City also recognized the need for water as essential for the growing the capital city of their state. Oklahoma City is about two hundred miles north of the Red River. In between these two cities, lies the home of the Chickasaw and Choctaw Nations. In the late 1950's, both metropolitan areas targeted southeast Oklahoma as the answer to their water needs. At the time, tribes did not possess the political influence to object or negotiate fair compensation for water pipeline easements through their property.

Oklahoma City "first participated in acquisition of Sardis Lake's water supply in the 1990's (Couch, 2007). In 1992, the Oklahoma Water Resources Board (OWRB) entered into a contract to sell the water to the North Texas Municipal Water District. In 1992–1993, the Oklahoma Department of Wildlife Conservation (ODWC) expressed environmental concerns (ODWC, 1992) and conducted impact studies regarding loss of wildlife habitat and bottomland hardwoods, waterfowl management units, and increased soil erosion.

Talks of selling water continued through the decade. In 1995, the Oklahoma Comprehensive Water Plan (OCWP) was updated to suggest formation of a permanent committee with Indian representation to address water issues (OWRB b, 1995).

By 1997, Oklahoma had defaulted on the Sardis payments to the Corps. Yet, the State was able to announce We've been able to deposit more money into the rainy-day fund than ever before. . . . Oklahoma's strong economic growth resulted in a $245.9 million deposit to the rainy day fund this year (Oklahoma ,1997). It is not clear why Oklahoma chose to go in default. Only in the future does it become clear that a plan would be laid for Oklahoma City to assume the debt—and the water. The battle of Sardis Lake has long exceeded the terms of the past three Governors for Oklahoma. In 1997, under Governor Frank Keating's administration, Oklahoma was notified by the Corps that the State was in default of payments. "The governor has been steadfast in his desire . . . and made it very clear to me very early on that he wanted to acquire the lake if we could resolve the title problems associated with the Indian claims" (Yeager, 1998, p. 1). Formal recommendations were made to include tribal interests in the planning process. Those recommendations were never implemented.

A new centennial arrived, and repeated recommendations are made. This time, in 2000, the OWRB submitted a Kiamichi River Basin report to the Oklahoma legislature and the recommendations were stridently familiar— develop a state and tribal water compact. The priority was to maintain water security for Oklahomans first and with "the highest priority afforded to those Oklahomans residing within the Kiamichi River Basin" (Farmer, 2000, p. 1).

In direct opposition to the OWRB stated recommendations with the "highest priority" being recognized for those within the Kiamichi River basin, Oklahoma City crafted their own master plan. "Acquiring water rights in Sardis Lake is an integral part of the 2003 Oklahoma City Master Plan (Couch, 2007, p. 1). "The engineering process is underway to develop plans and specifications" (Couch, 2007, p. 3).

Chickasaw Nation Governor Bill Anoatubby pens a cordial, yet formal, letter to then Oklahoma Governor Brad Henry. Anoatubby outlines his concern that no progress on the recommendation for a State and tribal alliance regarding water rights has occurred. In closing, the letter becomes more forceful and directs Henry's attention to the enclosure prepared by the tribal attorney with prescribed expectations for water dialogue (Anaotubby, 2008). The time had come to "string the bow" and be prepared to fight. Anaotubby's attached document is from Stephen H. Greetham, a water and natural resources attorney. The letter effectively represents the Chickasaw Nation and outlines the expectation for the State of Oklahoma and the OWRB that the Chickasaw Nation desires a collaborative relationship with the State and its agencies (Greetham, 2012).

Possibly pressured by Anaotubby's stern letter to the Governor, Oklahoma City appeared to acknowledge they would be unable to secure the water on their own volition. They recognized the increased legal position of the tribes' power and authority. Oklahoma City sought avenues to strengthen their own position. They joined forces with the surrounding central Oklahoma cities of Moore, Norman, Midwest City, Shawnee, Edmond, El Reno, Mustang, Yukon, and others. Their collaborative efforts funded the 2009 Regional Raw Water Supply Study for Central. "With existing resources fully utilized, it will be necessary to receive water from a new resource. . . . This resource is Lake Sardis. . . . This study investigates the best way to gain access to this resource and how best to deliver this water to the citizens of Central Oklahoma" (OCWUT, 2010). By that fall, a document was hand delivered to Oklahoma's State Treasurer, Scott Meacham. The document summarized the agreed upon terms for Oklahoma City to buy the Sardis Lake water that the State of Oklahoma was offering them for sale. It also noted that Oklahoma City "wishes to suggest amended terms" (White & Couch, 2009, p. 1). The salient point was Oklahoma City wanted the agreement to preclude that in any negotiations the State of Oklahoma made with Native American nations regarding Sardis Lake that "Oklahoma will neither diminish the rights and interests conveyed to Oklahoma City hereunder nor obligate Oklahoma City to further compensation or consideration for such rights and interests" (White & Couch, 2009, p. 2).

While the centrally located municipalities were collaborating to develop their own future water plan studies, the State of Oklahoma ordered a $14 million study for a comprehensive water analysis plan of their own. The study detailed the urgent needs of urban populations, farming, and oil and gas industry. The study failed to adequately address non-consumptive use of the water. Non-consumptive water use includes such activities as recreation and water related tourism. Even the Winters decision allocated water value in non-economic terms "The Indians had command of the lands and the waters—command of all their beneficial use, whether kept for hunting . . . or turned to agriculture and the arts of civilization" (Winters, 1908, p. 7). How the non-consumptive use of water is devalued through State ordered "research" resonates with the historical devaluing of sacred grounds and diminishing the true nature of tribal agrarian culture.

## Media Reaction: Non-Payment of Sardis Lake Debt

Tribal leaders were not the only injured parties in the plan. The report was released in April 2011 to strong criticism from other stakeholders as well. One legislator was disappointed to learn the expensive study did not even include basic quantified impacts of recreational and environmental uses of non-consumptive water. This was principally shameful since Oklahoma boasts a 7.1-billion-dollar economic impact from tourism (Oklahoma, 2013). Much of it focused in southeastern Oklahoma related to boating, fishing and other water related recreation. Legislators requested an Oklahoma Attorney General's opinion to determine "whether the OWRB has fulfilled its legislative mandate to make a comprehensive report since recreational and environmental uses were not quantified" (Adcock, 2011, p. 1). The millions of dollars of tax-payer funds spent for a document that contained such glaring omissions was yet another example of macroscopic governmental bureaucracy. It directly countered to the micro-subjective perceptions and beliefs of individuals that their values for recreational water, tourism and environmental respect were being ignored at best and omitted intentionally at worst. Furthermore, Jerry Ellis, the Oklahoma Senator that represented southeast Oklahoma, pointed out a direct conflict of interest for the private company involved with producing the study. "CDM engineering had contracts with the Comprehensive Water Plan and simultaneously worked for the Oklahoma City Water Trust authority in a contract to bring Sardis Lake water to Oklahoma City" (Ellis, 2011, p. 5). CDM Smith is a multinational corporation specializing in engineering for water and energy facilities. Their primary clients include governments, such as the Sardis water acquisition project for the State of Oklahoma and the City of Oklahoma City (Lewis, 2008).

The Oklahoma City water permit application for Sardis was looming on the immediate horizon. In April 2010, county commissioners for southeastern Oklahoma stood unified in direct opposition to the permit application. The commissioners presented formal resolutions of opposition to Oklahoma's water sale (Deela, Medders & Alford, 2010). "Texas doesn't want the water for residential use, but, rather, for natural gas development. It takes millions of gallons of water to hydraulically fracture, or frack, one natural gas well . . . there are more than 16,000 rigs in Fort Worth alone on the Barnett Shale" (Jacobs, 2013, p. 1).

On April 21, 2010 the leaders of both the Chickasaws and the Choctaws prepared their communication of protest. They sent another tough letter to Governor Henry. The tribes asserted they had been excluded from the closed negotiations related to Sardis Lake water. They introduced issues of actual State authority and jurisdiction. "Those are questions we would prefer to work out with you rather than resolve through formal conflict" (Carter, 2010, p. 1). The tone of the communication was pointed enough to put on ice the

$100-million-dollar deal for the OWRB and Oklahoma City to sell water (Carter, 2010, p. 1). The tribes were being rebuked at each attempt to assert their sovereign nation water rights. Legislators representing southeastern Oklahoma held a press conference on the marble steps leading to the Oklahoma capitol building decrying the secret agreement between the State of Oklahoma and Oklahoma City. The leaders renounced that the powerful people of Oklahoma City were exploiting the poverty of southeastern Oklahoma, one of the poorest areas of the state. Senators that represented the District requested to be included in any negotiations for water sales that were coming from the district. He "received no response whatsoever" from that request (Ellis & Wilson, 2010, p. 1). Following the press conference, the quiet deal in Oklahoma City was generating a lot of noise in the affected rural communities of southeastern Oklahoma. On Saturday, May 6, 2010, more than nine hundred concerned citizens attended a rally in Durant Oklahoma to further protest the water grab (Carter, 2010).

In addition to the citizens' rally garnering media attention, the continued non-payment of the debt on Sardis Lake was also being reported. With Oklahoma delinquent on the payments, interest continued to accrue. In 2009, the federal government filed a civil action to prompt a repayment with United States of America v. State of Oklahoma, et al. (2009). Oklahoma was ordered to pay almost $28 million dollars to cover indebtedness related to Sardis Lake. After one court ordered payment, they declared insufficient funds to make future payments. The State of Oklahoma sought a partner, "the only Oklahoma water supplier to put an offer on the table—Oklahoma City" (Lambert, 2010, p. 6).

Oklahoma's State Treasurer, Scott Meacham, calculated the costs, interests, penalties, future costs, and declared approval of the transfer and associated financial burden of Sardis Lake "equates to an over $270 million debt off the backs of Oklahoma taxpayers, while at the same time providing a valuable future water supply option to central Oklahoma communities" (Lambert, 2010, p. 7).

Meanwhile, the City of Oklahoma City and its agency, the Oklahoma City Water Utilities Trust (OCWUT) continued negotiations for water sales in the $100-million-dollar range (Carter 2010). The math is easy; Oklahoma City readily agreed to assume $28 million in debt. In October 2009, Oklahoma City courted the State of Oklahoma to accept their offer to assume the debt, pay the State of Oklahoma $15 million dollars, and the water would "ultimately be used to benefit several central Oklahoma communities" (Meacham, 2009, p. 2). Others argued Oklahoma City would "resell the water to western Oklahoma" (Muskogee, 2007, p.1). The water was not for the benefit of southeastern Oklahoma residents, nor for the Indian tribes that resided in the southeastern territory. Oklahoma City would benefit and arid western Oklahoma (Buchanan, 2013).

Plans of the agreement to transfer water rights and debt obligation from the State of Oklahoma to Oklahoma City came to the attention of the federal government. The Corps notified Oklahoma Governor Henry admonishing the actions between the State of Oklahoma and Oklahoma City, and advising "We find it prudent to remind you" of the federal government's role and authoritative rights for the management of Sardis Lake. Summarily, the State of Oklahoma and the City of Oklahoma City were not authorized to sell water that was not theirs to sell (Snyman, 2010, p. 1). This letter was followed shortly thereafter when the Chickasaw and Choctaw tribes collaborated and officially communicated to Governor Henry that they would make all the payments (Lambert, 2010). Oklahoma City was offering a buyout. The tribes' offer was more generous. They were willing to make the payments in order to maintain status quo—to buy another year. It would allow time to formulate a more comprehensive and inclusive solution. The OWRB acknowledged they were in receipt of that offer, yet it was declined. Additionally, information was available to the OWRB that the State had over $100 million remaining in the Rainy Day Fund, of which it was not out of reason that the $5.2 million payment in immediate demand could be made from, further staving the need for an immediate vote to approve Oklahoma City's water permit request. That option did not receive serious consideration. Lastly, the OWRB board could vote no, which in turn would replace the onus of responsibility for paying the debt back to the State legislature.

Notwithstanding the alternative options available to pay the debt, and in direct opposition to the warning advised from the Corps to the State of Oklahoma, a special meeting of OWRB was called for June 11, 2010—in order to approve the transfer of water rights to Oklahoma City. The meeting drew a large crowd that stood on both sides of the debate. State legislators from southeast Oklahoma, tribal nation representatives, lobbyists, county commissioners, and citizens attended the special meeting that seemed to be the culmination of Oklahoma City's fifty-year effort to acquire southeastern Oklahoma water. Chairman Rudolf J. Herrmann invited comments before the vote.

Jim Couch, City Manager of Oklahoma City, spoke for the pro-sale contingency of supporters. The greater Oklahoma City metropolitan represented about one-third of the entire state's population. They would need water for future growth. Couch referred to their similar water storage arrangements with Canton Lake and Atoka Lake and claimed their potential relationship with Sardis would be a "very similar relationship" in nature. What we do know is when Oklahoma City drew down Atoka Lake "when the water is taken there is nothing left but a mud hole" stated Greg Pyle, Chief of the Choctaw Nation (Lambert, 2010, p. 8). The taking of Sardis Lake water would discount the wants and needs of the oppressed population of residents in the basin of the Kiamichi Mountains. A series of newspaper advertisements from 1958 "hawked vacation lots along 'the scenic shores of Lake Atoka.' These ads implied hunting, fishing, boating, family fun and even inspiration would abound. Family fun was short-lived. By 1964, Oklahoma City had restricted access to the lake, and later fought to close the lake to fishing, boating and other recreational activities" (Embry, 2012, p. 5). In today's reality, Atoka Lake remains drawn down, fish habitats protrude from the dried lakebed, picnic tables are in disrepair. The real estate market is weak for lakeshore cabins on what was to be Atoka Lake.

With Atoka Lake viability incapacitated, Oklahoma City turned to their Canton Lake option. During extreme draught conditions, they took what they wanted, leaving the rural population of Canton without viable water sources for their own economic survival. Oklahoma City had been drawing down Canton Lake in previous years, but the draught of 2013 was portrayed as "What I call the 'kill shot' release, because this is the one that took the lake level to the point where it made the lake unusable" (Layden, 2013). During his 2010 testimony during the special session, Couch stated, "there is no need for this water [Sardis] for many decades and there will be a long time before . . . central Oklahoma has a need for the water" (Lambert, 2010, p. 7). Oklahoma City's assertation of water stewardship would become a deeply and bitterly tasting proposition for Atoka and Canton (Hennessy-Fiske, 2011). The OWRB touts "The State of Oklahoma has had great success protecting and allocating its citizens' water resources" (OWRB, 2012b, p. 4).

In closing, Couch offered "None of the water can go outside the State of Oklahoma" (Lambert, 2010, p. 7). However, what was known was that the city Hugo, a downstream location on the southern border of Oklahoma, would be an in-state permit holder and Hugo had already negotiated with North Texas permits. During the 2013 Tarrant v. Herrmann United States Supreme Court case, the cities of Hugo and Irving Texas together submitted to the Supreme Court justices an amici curiae (friend of the court) brief outlining their relationship and why the Court should find in favor of Tarrant. The City of Irving "a growing Texas municipality with projected water needs that far exceed current supplies, entered into a contract . . . to purchase water from Hugo. . . . Hugo currently holds water rights from the State of Oklahoma . . . and has a pending application to appropriate significant additional supplies" (Cottingham, Caroom, & Maxwell, 2013, p. 2). Couch stood before the committee claiming no out of state water permits would be allowed. He knew full well that permits to in state applicant, like that of Hugo, would be approved and Hugo would sell water to Texas.

With Couch's time before the OWRB board completed, State Senator Jerry Ellis was selected to present counter opinions on the matter of whether the OWRB should approve the Oklahoma City permit request for Sardis Lake. Senator Ellis represents the district where Sardis Lake is located. He avowed the negotiations to approve Oklahoma City taking the water were "cloaked in secrecy and lacks transparency" (Lambert, 2010, p. 9). Furthermore, the citizens of Sardis were forced to submit their own applications for water use and have had their applications waiting far longer than Oklahoma City had been waiting. In fact, the very point of this special session was solely to approve Oklahoma City's application. Two applications were submitted in 1993—and they remained languishing and unapproved by the Oklahoma Water Resources Board! "The OWRB loaded up the City of Hugo with water use permits to facilitate a Texas water deal" (Lambert, 2010, p. 9). Of all the OWRB members, not one resided in southeast Oklahoma, where the water is located.

If the facts were true as stated by Couch, that Oklahoma City would not actually need to the water f or decades, why the rush for a special meeting to approve the water storage transfer to Oklahoma City? Another southeastern legislator reminded the board that Oklahoma statutes required appraisals for items being sold. Items valued at over one million dollars could not be sold for less than ninety percent of the value. He ascertained the true value of the water had not been satisfactorily identified. Further opposition to the vote was voiced citing the Texas A&M research that indicated when transferring water from a basin "so goes with it the economic development, tourism, recreation, and the donor basin is left with environmental degradation" (Lambert, 2010, p. 11).

Following remarks by Senator Ellis, State Representative Brian Renegar affirmed that during the 2008 legislative session, $66 million in bond issues were being approved. Several legislators asked that funds for Sardis be included and they were told "that has already been worked out" (Lambert, 2010, p. 11).

The pressure to restrain the State of Oklahoma and the City of Oklahoma City from taking the water was relentless. On the same day of the special meeting, June 11, 2010, the U.S. Department of Interior joined the conflict. Larry Echo Hawk, Assistant Secretary of Indian Affairs weighed in by requesting the members of the Oklahoma Water Resource Board delay action on the transfer of Sardis water until appropriate consultations had been held with the Chickasaw and Choctaw Nations and federal officials (Allen, 2010).

## Water Board Forces Issue

The special meeting of the OWRB was drawing to a conclusion. Chairman Herrmann ruled adequate opportunity for comment had occurred. He called for a vote to approve the agreement between the Oklahoma City Water Utilities Trust and the Oklahoma Water Resources Board relating to the transfer of Sardis Lake water. The motion was approved, five voted to allow the water transfer and two voted no. The meeting adjourned at 11:30 a.m. It was a long four drive home for the citizens of southeast Oklahoma who had·

travelled to Oklahoma City in hopes their voices might be heard. However, there were a number of entities listening to the volleys being hurled back and forth, and not everyone approved of what they were hearing.

The tense special meeting ended and the OWRB wasted no time in executing the contract. On June fifteenth, the agreement was signed. And, on June thirtieth the OWRB sent a copy of the agreement to the U.S. Army Corps of Engineers requesting their approval—as was requested in their communication of May twentieth to the OWRB. "I am sure you have been kept apprised of the informal discussions over the past several months (emphasis mine) between our respective staffs regarding what we understand is a fairly perfunctory approval process" (Roselle, 2010, p. 1). The letter intones the public comments entertained at the June tenth special meeting were never sincerely considered. The deal had been sealed months beforehand.

With pressure building, the Corps did not "fairly perfunctorily" approve the transfer agreement. Instead, they referred the request to their District Counsel. Their response to the OWRB reaffirmed the Corps position of power. "A decision on an approval of a transfer and assignment will be made by the Assistant Secretary of the Army (Civil Works). . . . The decision-making process will take into account all available information to determine whether approval of a transfer is in the best interests of the United States" (Roselle Jr, 2010, p. 1). An additional hurdle presented to the OWRB was that before the transfer could occur, the entity (Oklahoma City and the OC-WUT) to which the contract is to be transferred to must have already obtained a valid water rights permit. And the Corps was "aware of other entities' claims to water rights at Sardis Lake that have been well-publicized in the local media . . . the Corps of Engineers does not make determination of water rights and does not become a party to disputes regarding water rights" (Roselle Jr, 2010, p. 2). Clearly, the Corps had been informed exactly how long some water permit applications had been languishing at the OWRB far ahead of Oklahoma City's permit requests.

Two additional reasons for avoiding public discourse and transparency were identified. The effort to avoid scrutiny of the transaction, even from residents of Oklahoma City and the surrounding towns, was because they would bear the financial risks for hundreds of millions of dollars of construction bonds for the project. Secondly, state-wide controversy existed on the possibility of Texas receiving the water. It was possible the central Oklahoma residents would pay for construction to bring the water from southeast Oklahoma to central Oklahoma for what they believed would be for their benefit, only to learn the water could then be transported away to Texas. This possibility was ever at the forefront of discussion and all eyes were upon Texas as *Tarrant v. Herrmann* was consistently in the media as it headed closer and closer to the U.S. Supreme Court.

METHODOLOGY

## Data and Data Analysis

Data was coded in terms of document type (e.g. formal letter, informal communication, government document, tribal document). Levels of power and authority were categorized as high, medium or low and identified as formal, informal, and charismatic. For instance, a letter from Ken Salazar, United States Secretary of the Interior was coded as high formal authority. "States have been the world's largest and most powerful organizations" (Tilly, 2012, p. 252). The strength of a nation-state is predicated on their ability to maintain adjudication, distribution, production and extraction. In other words, the successful state will aim to maintain authority for settling disputes; intervene in the allocation of goods amongst the members, control goods and services, and draw out resources of the subject population (Tilly, 2012). Data was further coded and analyzed for negative descriptors (e.g. exploitive, coercive, domineering) and positive (e.g. facilitate concern, solve, satisfy, recommend).

Content and narrative analysis provide the rich and thick descriptions commonly expected in qualitative inquiry. It permits for the study of processes over time and it is an unobtrusive method that has little effect on human subjects. Additionally, the concreteness of the documents strengthens reliability (Babbie, 2011). Consideration was given to macroscopic and microscopic elements of the conflict, as well as consideration toward objective and subjective conditions. Lastly, Ritzer's theoretical model demonstrates the intersectionality of complex components and their influences on each other (Ritzer, 1991; Ritzer, 1995).

A broad review of historical context is presented. I t may appear redundant to extensively revisit the past when the research question is centered upon issues related to the current conflict over Sardis Lake water. However, "The focused revisit takes . . . on very different meanings because of changes in historical context and the interests and perspectives of the revisitor" (Burawoy, 2003, p. 650). The review of historical land rights and the underlying water rights are insightful tools for comparisons when evaluating today's dilemma of water rights and transactions. The contextual elements of the present case become clearer with the specificity for the past. This illumination is partly clarified from the fact that much of the earlier writings regarding land transactions were "conducted under the protective guardianship of colonialism—conditions that remained silent in the original studies" (Burawoy, 2003, p. 649). This position is reaffirmed by Native American scholars that characterize studies by American anthropologists as "ethnocentric and implicitly colonialist" (Erickson, 2011, p. 49). Following the goal for templates of social justice inquiry, the analysis placed "the voice of the oppressed at the center of inquiry" (Denzin, 2010, p. 103). The importance of this perspective is that the privileged versions "relies upon a substructure that has already discredited and deprived of authority to speak the voices of those who know the society differently" (Smith, 2012, p. 404). By beginning with the historical overview, voices previously unheard are quite clear.

Underlying circumstances of the water conflict for Sardis Lake rest in historical rights afforded to the tribes. The United States government federally recognizes 566 tribes (NCSL 2014). Of those, the so-called Five Civilized Tribes of Oklahoma (Cherokee, Chickasaw, Choctaw, Muscogee-Creek, and Seminole) expressly have the benefit of additional rights in their claims to land and natural resources—including water rights.

## DISCUSSION

### Letter to Another Governor

Dissatisfied with stalled efforts for inclusion, on August 18, 2011, the Chickasaw and Choctaw tribes wrote another letter to another Governor. This time they addressed their continued concerns to Oklahoma Governor Mary Fallin "State law *does not* control this subject [of water]. . . . We expect that our continued forbearance would simply mean the deepening of our present challenges. . . . History and the law demonstrate that we must act" (Anoatubby & Pyle, 2011, p. 1). The letter reproved Oklahoma's handling of the water issue by their recitations of state laws—which the tribes iterated are in an inferior position to federal law—where the water conflict must be ruled from. Their 36-page lawsuit, Chickasaw v. Fallin was filed on August 18, 2011 in the U.S. District Court for the Western District of Oklahoma. The lawsuit alleges the Indian Nations have federally protected rights to the water within a territory in southeastern Oklahoma that are prior and paramount to any rights granted by the State to Oklahoma's citizens. The lawsuit primarily sought to halt any permit application action by the OWRB for Oklahoma City to use Sardis water or from exporting water until general stream adjudication has been satisfied (Chickasaw, 2011).

With fewer options remaining, in February 2012, Oklahoma quickly assumed the position to oversee the stream adjudication process that was going to be required. "The State, through the OWRB is authorized to commence a general stream adjudication . . . to confirm and determine rights" (OWRB, 2012b, p. 2). In response to the Chickasaw v. Fallin action, the OWRB, acting under the authority of the State of Oklahoma, filed a lawsuit in state court, Oklahoma Water Resources Board v. United States on behalf of the Choctaw Nation of Oklahoma, a federally recognized Indian Tribe; the United States on behalf of the Chickasaw Nation, etc., et al. (OWRB v. United States, 2012). This action was designed to force the state stream adjudication process. For a state, such as Oklahoma, to instigate the water rights adjudication process, the McCarran Amendment was intended to allow the United·States to be enjoined as a defendant for no other reason than "Unless all the parties owning . . . water rights . . . can be joined as parties any subsequent decree would be of little value" (McCallister, 1976, p. 305).

Before the Oklahoma water rights adjudication could take place, the federal government removed the lawsuit for adjudication from state to federal court (Coats et al., 2012). By prompting the McCarran adjudications, Oklahoma impelled the federal government, at the highest levels, to become involved. The Secretary of the Interior, Ken Salazar, explained the immediate concern and process that would be implemented regarding the adjudication. The determination of water rights would not be expeditiously decided by Oklahoma. "If, in the end, comprehensive water rights adjudication must take place, we would . . . discuss the proper forum for conducting the adjudication" (Salazar, 2012, p. 1).

In March of 2012, federal court mediator Frances McGovern issued an order to stay formal proceedings (put the case on hold) for sixty days in Chickasaw v. Fallin to allow more time for mediation among the parties. Oklahoma and the OWRB were particularly attuned to their other lawsuit against them, Tarrant v. Herrmann. Oklahoma had beat Texas on every level of the jurisdictional process and through every appeal. Texas, undeterred, and spending millions of dollars in legal fees, requested a U.S. Supreme Court hearing. After agreeing to hear the case, the Supreme Court justices invited U.S. Solicitor General (SG), Donald B. Verrilli, Jr. to submit his opinion on the matter. Often referred to as the tenth justice because of the weight afforded by the justices to the SG's opinion, Verrilli, Jr. submitted "Water rights of the Tribes [Chickasaw and Choctaw] may be relevant to the amount of excess water available" (Verrilli, Jr., 2012, p. 20). With the SG's respected opinion on the record, the tribes had good reason to be hopeful. Although the tribes were not directly a party to Tarrant v. Herrmann, they watched the case closely. The tribes even provided their own friend of the court brief in support of Oklahoma. Previously, Tarrant had attempted to buy water from the Choctaws and Chickasaws. This attempt was unsuccessful (Sotomayor, 2013). When negotiations for purchase failed, the tactic was amended. In 2009, Tarrant attempted to buy Red River water from the Apache Tribe of Oklahoma and this too was unsuccessful (Chalepah & Oliver, 2009).

## U.S. Supreme Court Hears Water Dispute

The day arrived. Tarrant v. Herrmann was heard on April 23, 2013 by the United States Supreme Court. If Oklahoma prevailed, it could be considered a strength to Oklahoma and/or Indian water rights. Representing Herrmann (Oklahoma) for the oral arguments was Lisa Blatt. She "has argued 33 times before the U.S. Supreme Court. She's won 32 of those times" (Carter 2013). Her oral arguments were persuasive, but would they be enough to win? While the justices deliberated, the parties to *Chickasaw v. Fallin* waited to hear how Tarrant v. Herrmann would be adjudicated. And they waited. The court had not rendered an opinion by May 2012 when the Chickasaw v. Fallin case received an extended stay for another 60 days. The parties agreed they were "making progress" in negotiations and both parties affirmed it was in the best interest for judicial proceedings to be "stayed" so discussions could continue. During the stay, news arrived that the U.S. Supreme Court had rendered a decision for Tarrant v. Herrmann. In June 2012, the Supreme Court announced their verdict on *Tarrant*. The court ruling was unequivocal and the message from the Justices was clear. In a unanimous opinion, all nine U.S. Supreme Court justices ruled in favor of Oklahoma. Texas could not take their water. Blatt won. It was a victory for southeastern Oklahoma. The opinion rested on the element that "Adopting Tarrant's reading would necessarily entail assuming that Oklahoma . . . silently surrendered substantial control over their waters" (Sotomayor 2013:3). For their case, Chickasaw v. Fallin, the tribes were also arguing they had never surrendered their waters. The *Tarrant* ruling appeared to favor the tribes' position. Power was shifting. Negotiations granted through the stays for Chickasaw v. Fallin now would reflect that new knowledge available from the Tarrant opinion. The State of Oklahoma was more likely to respect the process of negotiation over litigation with tribes for Chickasaw v. Fallin.

## Power and Influence

Scrambling to secure additional power and influence, in August 2013, Oklahoma Governor Fallin solidified the state's liaison with the powerful and well-funded Corps. Immediately following his retirement from the Corps, Michael Teague joined Fallin's organization as Secretary of Energy and Environment, a newly created cabinet level position. "His years of experience dealing with . . . water management will serve him well as Oklahoma's first secretary of energy and environment" (Fallin, 2013, p. 1). Adding this player to the roster for Oklahoma has been the final known act in this battle for power and control over Sardis Lake water. Federal Judge Lee West issued a gag order restricting the parties from disclosing further details of the ongoing negotiations. West, overseeing the negotiations is "likely to keep issuing stays if the state and tribes keep requesting them" (Wertz, 2013, p. 1).

The State of Oklahoma and its agencies, including the OWRB have asserted authority for the Sardis Lake water. They have negotiated to sell the water out of the basin of origin to the city of Oklahoma City. The water would be transported via pipeline to benefit Oklahoma City and the surrounding urban populations of central Oklahoma. Opinions have been voiced that the State of Oklahoma and the city of Oklahoma City further intend to construct pipelines to transport the water to western Oklahoma and to sell the water to urban populations in the Dallas-Fort Worth Texas region (Buchanan, 2013). At an annual Oklahoma Governor's water conference, a presentation "We're all in this together" identified the collective needs for water users. Oklahoma City and surrounding metropolitan communities were identified as "stakeholders". Tribes were labelled as "affected party" (Lewis, 2008, p. 5).

The actions for taking Sardis Lake waters were a familiar scenario for the tribes. It was redolent of the "negotiations" for leaving their Mississippi homelands. The next generation experienced the railroad right of way constructions through their Oklahoma lands in the1880s. These actions had far reaching economic, environmental, social, and cultural impacts. The history of fair negotiation between the government and the tribes was not positive. . The more recent exploitation regarding the Atoka Lake pipelines being built through their lands without fair compensation was a poor meter for imagining how the Sardis Lake conflict would be managed.

The tribes have been steadily improving their levels of education, business acumen, and legal understandings of their rights. They have armed themselves with expert advisors. They chose to defend their water rights to Sardis Lake. This case study revealed how the turn of events from historical and current power structures and conflicts affected the attempted sale of water from Sardis Lake. The government actors underestimated the will and the power of the tribes.

As power shifted in the battle of the Sardis Lake water sale, efforts toward mediation, rather than legal recourse, were improved. Reviewing how land rights evolved presented a platform for better understanding how water rights can be negotiated and the context from which the parties entered into the debates. For Sardis Lake water rights, it can be asserted that the Chickasaw and Choctaw tribes have the benefits of sovereign rights and ownership rights. It is also clear federal law, not state, will govern the direction of water authority (Greetham, 2012).

## Conflicts over Sardis

When conflicts over Sardis water escalated, it was discovered that when powers shifted, efforts toward mediation rather than legal recourse were improved. Hence, both sides of Chickasaw v. Fallin acknowledge they are making progress toward negotiation, staying off expensive and time-consuming litigation. Indians were divested of their lands through a combination of threats, thievery and trickery. Stealing the waters will be a tougher battle. Many water claimants are recognizing friendlier and fairer negotiations may well be the avenue of choice. The Big Horn stream adjudication in Wyoming was a protracted affair. The case languished in the court system for more than thirty years. The involved parties spent over thirty million dollars (Miller, 2012). The State of New Mexico v. Aamodt case began in 1966 and has been considered one of the longest running cases in federal judicial history. The case mired more than forty years of federal court resources. The case was settled in 2010 and in 2013 the settlement documents were actually signed (Kershaw & Darling, 2013). An estimated 200-million-dollar price tag has been associated with the suit and fulfilling the conditions of the settlement, are expected to be completed by 2017, a full 51 years after the suit was filed (Mathews, 2013). This process of adjudication can be an expensive, protracted, and emotional for all parties involved.

## CONCLUSION

The federal government's involvement in Indian water rights cases emanates from treaty doctrines. Historically, it has been demonstrated that the executive branch has repeatedly failed that promise. The Reagan administration began the current policy of making an effort to avoid adversarial litigations and attempt amicable negotiations of tribal water rights. The U.S. Department of Interior and Army Corps of Engineers involvement in the Sardis case reflects this current federal position.

How Chickasaw v. Fallin was ultimately negotiated when plaintiffs and defendants resolved their conflict through peaceful negotiation, not litigation. The tribes are a people who look forward and not at "the bloody footprints behind them" (Tingle, 2003, p. 51). They eagerly embrace the opportunity to sit justly at the table of dialogue for "protecting and preserving the sustainability of water . . . that is fair, meaningful, and serves the best interest of all Oklahomans" (Choctaw, 2011, p. 1).

In the final negotiated settlement of 2016, the Oklahoma Indian tribes did not receive any financial compensation for allowing the State of Oklahoma and the city of Oklahoma City to take the water from Sardis Lake. What the tribes did receive, was a promise that a specific amount of water would be reserved in the reservoir, Sardis Lake, for local use. Additionally, water diversion from the Kiamichi River is required to meet specified minimum flow rates. The Kiamichi River shall not be drained dry. The agreement was established to protect fish and wildlife as well as downstream users in the rural community of Antlers Oklahoma.

The State of Oklahoma and the city of Oklahoma City each allocated over two million dollars of their taxpayer money to litigate the tribes for water rights. Unlike with the Indian treaties of the 1830s, the tribes maintain they now have the resources to strictly enforce the conditions of the agreement. Choctaw Chief Gary Batton iterated "Due to our efforts with congress, the regulatory provisions secured by this settlement agreement are now enforce-able in federal court. Not only does this protect our tribal sovereignty, it also provides a fair playing field for any future disagreements" (Turner, 2019, p. 1). President Barack Obama signed the 2016 Water Infrastructure Improve-ments for the Nation Act, which was crafted based on the southeastern Okla-homa dispute. The intent is to lessen water ownership disputes and create a context for future collaborations between federal, state, local, and tribal en-tities (Brus, 2019).

Although the lawsuit has been settled out of court by a negotiated agree-ment amongst the parties, litigation surrounding the agreement continues. The Kiamichi River Legacy Association (KRLA) argues conditions of the agreement could dramatically alter the Kiamichi River's flow and water level. These changes could violate the Endangered Species Act by harming indigenous mussels. The Ouachita rock pocketbook and the scaleshale are federally protected. With these violations in mind, the group filed a "brief-in-chief" petition in Oklahoma's Pushmataha County District Court (Wertz, 2019). Additionally, in April 2019, the KRLA with other plaintiffs, filed a lawsuit in Muskogee's United States District Court for the Eastern District of Oklahoma. The defendants include Oklahoma Governor Kevin Stitt; the ex-ecutive director of the Oklahoma Water Resources Board (OWRB); Oklaho-ma City Mayor David Holt and Carl Edwards, chairman of the board of trustees of the Oklahoma City Water Utilities Trust; Bill Anoatubby, govern-or of the Chickasaw Nation, and Gary Batton, Chief of the Choctaw Nation of Oklahoma; plus the acting Secretary of the U.S. Department of the Interi-or, as Trustee of the Choctaw and Chickasaw Nations (Ray, 2019).

## BIBLIOGRAPHY

Adcock, C. (2011). Climate change and population increase could cause possible water short-ages. *Oklahoma Gazette*. Retrieved from http://okgazette.com/oklahoma/article-12526-h-2-no.html.

Anoatubby, B., & Gregory, P. (2011, Aug 18). Chickasaw Nation, Choctaw Nation legal action filed today. Letter to Governor Mary Fallin. Retrieved from http://s3.amazonaws.com/choc-taw-msldigital/assets/ 1250/legalactionfiled_original.pdf.

Anoatubby, B. (2008). Letter to Governor Brad Henry regarding a proposal for water dialogue. Retrieved from http://s3.amazonaws.com/choctaw-msldigital/assets/1248/anoatubby letter_original.pdf.

Babbie, E. (2011). *The basics of social research*. Belmont CA: Wadsworth.

Brus, B. (2019). Long dispute over Sardis Lake water rights settled. *The Journal Record*. Retrieved from https://journalrecord.com/2019/02/26/long-dispute-over-sardis-lake-water-rights-settled/ .

Buchanan, T. (2013). Audio-Altus municipal trust authority/city council meeting. City of Altus. Retrieved from http://www.cityofaltus.org/shell.asp?pg=214.

Burawoy, M. (2003). Revisits: An outline of a theory of reflexive ethnography. *American Sociological Review*, 68 (5), 645–79.

Burrage, M., Rabon, B. & Greetham. H. S. (2011). Chickasaw Nation and Choctaw Nation of Oklahoma v. Fallin et al. Retrieved from http://www.owrb.ok.gov/util/pdf_util/lawsuitdocs/ Complaint_ChickChoc_08-18-11.pdf.

Canby, Jr. W. C. (1988). *American Indian law in a nutshell.* St. Paul MN: West Publishing Co.

Carter, M. S. (2013). From Veronica to Tarrant: High-powered attorney familiar with Oklahoma. *Journal Record.* Retrieved from http://journalrecord.com/2013/11/20/from-veronica-to-tarrant-high-powered-attorney-familiar-with-oklahoma-law/#ixzz2rj1DSf4S.

Carter, M. S. (2010). Troubled Waters—Sale of Sardis Lake put on hold. *Journal Record.* Retrieved from http://journalrecord.com/2010/05/10/troubled-waters-capitol/.

Chalepah, E. A., & Oliver, J. (2009). Memorandum of understanding. Anadarko OK: Apache Tribe of Oklahoma. Retrieved from http://s3.amazonaws.com/content.newsok.com/documents/Exhibit2ApacheMOU.pdf.

Choctaw Nation. (2011). Chickasaw and Choctaw Nations file lawsuit to protect water rights. Choctaw Nation Press Room. Retrieved from http://www.choctawnation.com/news-room/ press-room/media-releases/chickasaw-and-choctaw-nations-file-lawsuit-to-protect-water-rights/.

Coats, C. S., Sewell, K., Moreno, S. I., & Sprague G. M. (2012). Notice of removal to the United States District Court for the Western District of Oklahoma. U.S. Department of Justice. Retrieved from http://www.owrb.ok.gov/util/pdf_util/lawsuitdocs/Notice-of-Removal.pdf.

Cosens, B., & Royster, V. J. (2012). *The Future of Indian and federal reserved water rights: The winters centennial.* Albuquerque, NM: University of New Mexico Press.

Cottingham, E. D., Caroom G. D., & Maxwell, M. S. (2013). *Brief of Amici Curiae City of Irving, Texas, City of Hugo, Oklahoma, and Hugo Municipal Authority in Support of Petitioner* in *Tarrant regional water district v. Herrmann et al.* Supreme Court of the United States. Retrieved from http://www.Americanbar.Org/Content/Dam/Aba/Publications/Supreme_Court_Preview/Briefs-V2/11-889_Pet_Amcu_Coi-Etal.Authcheckdam.Pdf.

Couch, D. J. (2007). Council agenda item no. VI.CC. 3/27/07. The City of Oklahoma City. Retrieved from http://www.okc.gov/council/council_library/packet/070327/VI%20CC.pdf.

Deela, D., Medders, J. & Alford, R. (2010, Apr 5). Resolution. Latimer County Commissioners. Retrieved from https://www.sai.ok.gov/Search%20Reports/database/Latimer-CoOp12WebFinal.pdf.

Denzin, K. N. (2010). *The qualitative manifesto.* Left Coast Press: Walnut Creek CA.

Ellis, J. (2011). Lawmakers seek AG's opinion on water study. *OWRP Times* 1(2).

Ellis, J. & Wilson, J. (2010). Southeastern Oklahoma lawmakers decry secret grab for control of Sardis Lake. Oklahoma State Senate press release. Retrieved from http:// www.oksenate.gov/news/press_releases/press_releases_2010/pr20100420dpv.html.

Embry, P. (2012). Stewardship v. a pipeline. *Oklahoma Water Issues,* 2(2), 4–5.

Erickson, F. (2011). A History of qualitative inquiry in social and educational research. In N. K. Denzin and S. Yvonna S Editor (Eds.), *The sage handbook of qualitative analysis (42–59),* Lincoln. Los Angeles: Sage.

Fallin, M. (2013). Governor Mary Fallin appoints new secretary of energy and environment. State of Oklahoma press release. Retrieved from http://services.ok.gov/triton/modules/newsroom/newsroom_article.php?id=223&articleid=12432).

Farmer, L. (2000). Untitled. Correspondence to 48th Oklahoma State legislature. Retrieved from Oklahoma City OK: State of Oklahoma Water Resources Board.

Greetham, H. S. (2012). Symposium: Oklahoma's 21st century water challenges: Water planning: an opportunity for managing uncertainties at the tribal-state interface? *Oklahoma Law Review,* 64(593), 14.

Henderson, T. (2011). Five tribes' water rights: Examining the Aamodt adjudications' Mechem doctrine to predict tribal water rights litigation outcomes in Oklahoma. *American Indian Law Review,* 36 (1), 125–160.

Hennessy-Fiske, H. (2011). Tribes, small-town residents fear Oklahoma City will drain their lake. *Los Angeles Times.* Retrieved from http://articles.latimes.com/2011/nov/13/nation/la-na-oklahoma-lake-20111113.

Jacobs, P. J. (2013). Supreme Court wades into bitter Texas-Okla. feud ahead of expected 'flood of litigation. *Environmental and Energy Publishing*. Retrieved from http://www.eenews.net/stories/1059977696 .

Judy, A. (2010, August 14). Federal Indian advisors pay visit to Sardis Lake. Retrieved from https://www.mcalesternews.com/news/federal-indian-advisors-pay-visit-to-sardis-lake/article_462bedfb-4905-5dec-9d07-cf0e7cd57d11.html .

Kershaw, j., & darling, n. (2013). water settlement one of six settlements reached during the Obama Administration that will help deliver clean drinking water, certainty to water users across the West. Retrieved from the U.S. Department of Interior.

Lambert, L. (2010). Special meeting official minutes. Oklahoma Water Resources Board. Retrieved from http://www.owrb.ok.gov/news/meetings/board/board_pdf/2010/bdminutes_0610spec.pdf.

Layden, L. (2013). Canton businesses on the brink months after Oklahoma City water withdrawal. National Public Radio member stations. Retrieved from http://stateimpact.npr.org/oklahoma/2013/11/07/canton-businessowners-on-the-brink-months-after-oklahoma-city-water-withdrawl/.

Lewis, S. A. (1992). Letter to OWRB from ODWC. Retrieved from http://s3.amazonaws.com/choctaw-msldigital/assets/1246/ odwcsardis_ original.pdf .

Lewis, S. (2008). Oklahoma regional water supply infrastructure study. Paper presented at the Mathews, K. (2013). Aamodt settlement finally signed but not yet delivered. *Santa Fe Water Awareness Group*. Retrieved from http://waterawarenessgroup.wordpress.com/2013/04/17/aamodt-settlement/ ).

Mathews, K. (2013). Aamodt settlement finally signed but not yet delivered. *Santa Fe Water Awareness Group*. Retrieved from http://waterawarenessgroup.wordpress.com/2013/04/17/aamodt-settlement/ .

McCallister, E. (1976). Water Rights: The McCarren Amendment and Indian Tribes' Reserved Water Rights. *American Indian Law Review*, 4(2), 303–310.

McCool, D. (2002). *Native Waters: Contemporary Indian Settlements and the Second Treaty Era*. Tucson AZ: University of Arizona Press.

Meacham, S. (2009, Nov 13). Sardis Lake water. Letter to James Couch, City Manager of Oklahoma City. Retrieved from http://www.orwp.net/wp-content/uploads/2012/03/JDSltrtoCOE63010.pdf

Miller, J. R. (2012). *Reservation capitalism*. Lincoln NE: University of Nebraska Press.

Mollinga, P.P. (2008). Water, politics and development: Framing a political sociology of water resources management. *Water Alternatives*, 1(1), 7-23.

Muskogee P. (2007). Settle Sardis Lake debt. *Muskogee Phoenix*. Retrieved from http://www.muskogeephoenix.com/editorials/x2128952877/Settle-Sardis-Lake-debt/print.

National Conference of State Legislatures (NCSL). (2014). Federal and state recognized tribes. *National Council of State Legislatures*. Retrieved from http://www.ncsl.org/research/state-tribal-institute/list-of-federal-and-state-recognized-tribes.aspx#ok .

Oklahoma City Water Utilities Trust (OCWUT) (2010). Regional raw water supply study for central Oklahoma. Prepared by CDM Smith. Retrieved from http://www.okc.gov/agenda-pub/cache/2/snq1fd3ymwirihbsw3cxbz55/86593201252014012631924.PDF.

Oklahoma State Senate. (1997). Legislator wants to expand rainy day fund . . . Press release dated September 9, 1997. Retrieved from http://www.oksenate. gov news/press_releases/press_releases_1997/PR970909.html.

Oklahoma Water Resources Board (OWRB). (1995). Update of the Oklahoma comprehensive water plan. Retrieved from http://www.owrb.ok.gov/supply/ocwp/ocwp1995.php .

Oklahoma Water Resources Board (OWRB). (2012). Protecting Oklahoma's water resources. Retrieved from http://www.owrb.ok.gov/util/pdf_util /lawsuitdocs/GeneralStreamAdjudication.pdf.

Ray, M. (2019). Lawsuit to block OKC from siphoning water from Kiamichi River and Sardis Lake filed in federal district court. Retrieved from https://www.reddirtreport.com/red-dirt-news/lawsuit-block-okc-siphoning-water-kiamichi-river-and-sardis-lake-filed-federal .

Ritzer, G. (1991). *Frontiers of social theory: The new syntheses*. NY: Columbia University Press.

Ritzer, G. (1995). *Expressing America: A critique of the global credit card society.* Thousand Oaks CA: Sage.

Robertson, G. L. (2011). Oklahoma comprehensive water plan supplemental report: tribal water issues & recommendations. *Oklahoma Water Resources Board.* Retrieved from http://www.owrb.ok.gov/supply/ocwp/pdf_ ocwp/ WaterPlanUpdate/draftreports/OCWPTribal-Water_IssuesRecs.pdf.

Roselle, Jr., J. (2010). Letter to J.D. Strong. Retrieved from http://www.orwp.net/wp-content/uploads/2012/03/USACEreplytoOWRB71310.pdf .

Salazar, K. (2012). Letter to Governor Fallin. Retrieved from http://www.owrb.ok.gov/util/pdf_util/lawsuitdocs/DOI-letter-GovFallin_03-19-12.pdf .

Smith, E. D. (2012). Coercion, Capital, and European states [1990] In *Contemporary Sociological Theory*, edited by Craig Calhoun, Joseph Gerteis, James Moody, Steven Pfaff, and Indermohan Virk. Wiley-Blackwell: Malden MA.

Snyman, E. (2010). Transfer of Sardis Lake Water Storage. Letter to Honorable Brad Henry, Governor, State of Oklahoma. Retrieved from http://www.okwaterlaw.com/Portals/32/USAGE_Ltr_to_Governor_re_Sardis_20May10.pdf .

Sotomayor, J. (2013, Apr 23). Tarrant regional water dist. V. Herrmann Opinion Syllabus 569 U. Retrieved from https://supreme.justia.com/cases/federal/us/569/614/.

Tilly, C. (2012). Coercion, capital, and European states [1990]. In C. Calhoun, J. Gerteis, J. Moody, S. Pfaff, and I. Virk (Eds.). *Contemporary sociological theory* (252–255). Wiley-Blackwell: Malden MA.

Tingle, T. (2003). *Walking the Choctaw Road.* El Paso TX: Cinco Puntos Press.

Turner, K. (2017). Revisiting the water settlement for Sardis Lake. *Choctaw Nation of Oklahoma.* Retrieved from https://www.choctawnation.com/chief-voice/revisiting-water-settlement-sardis-lake.

U.S. Army Corps of Engineers (USACE). (2019). Welcome to Sardis Lake. Retrieved from https://www.swt.usace.army.mil/Locations/Tulsa-District-Lakes/Oklahoma/Sardis-Lake/ .

U.S. Department of Interior. (2013). What does tribal sovereignty mean to American Indians and Alaska Natives? *Bureau of Indian Affairs.* Retrieved from http://www.bia.gov/FAQs/index.htm .

Verrilli, Jr. B. D. (2012). *Brief for the United States as Amicus Curiae.* Washington: Supreme Court of the United States, Department of Justice No. 11-889: 1–22.

Wertz, J. (2013). State and tribes still wrestling over water rights in Oklahoma. National Public Radio member station. Retrieved from http://stateimpact.npr.org/oklahoma/2013/07/08/state-and-tribes-still-wrestling-over-water-rights-in-oklahoma/.

White, P., & James. D. Couch, D. J. (2009, Oct 21). Untitled. Letter to Scott Meacham, Oklahoma State Treasurer. Retrieved from http://www.orwp.net/wp-content/uploads/2012/02/Jim-Couch-Letter-10-21-09.pdf.

Winters v. The United States. (1908). No. 158. Supreme Court of the United States. 207 U.S. 564; 28 S. Ct.207; 52 L. Ed. 340; 1908 U.S. Lexis 1415. Retrieved from http://resources.utulsa.edu/law/classes/rice/USSCT_Cases/Winters_v_US_207_564.htm .

Yeager, J. (2014). State defaults on $6.3 million Sardis Lake debt to Government. NewsOk, Retrieved from http://newsok.com/state-defaults-on-6.3-million-sardis-lake-debt-to-government/article/2618741 .

Yin, K. R. (2014). *Case study research: design and methods.* Los Angeles CA: Sage.

# Index

# Author Biographies

**Dr. Dorothy Greene Jackson** is professor of nursing at the University of Texas Permian Basin. Dr. Jackson holds a Ph.D. in Nursing Science from Texas Woman's University; Master of Science and Community Health-Family Nurse Practitioner from the University of Texas at El Paso and Master of Science in Adult Health from the University of Texas Medical Branch, Galveston Texas. Dr. Jackson was the founding Dean of the College of Nursing at insert school name, where she launched the Bachelor of Science and RN to BSN programs, and the initial national accreditation of the nursing school. She currently serves on the cardiovascular team for the American College of Cardiology and was a recent member of the Cardiovascular Training Section Leadership Council. Her major interests are health disparities in heart disease, environmental, community and public health.

**Dr. Dian Jordan** is a senior lecturer for University of Texas Permian Basin. She received her Ph.D. from Texas Woman's University and Master of Human Relations from the University of Oklahoma. Dr. Jordan has previously worked for 500 companies and assisted with "Made in Oklahoma" agricultural products international sales and export incentives.

She serves on the boards for Southeastern Oklahoma State University-McCurtain County Campus and the McCurtain County Excise and Equalization. She has published articles in *Qualitative Methods*, *Contemporary Sociology*, and *Race, Gender & Class* peer-reviewed journals. Dr. Jordan's service to profession includes serving as a manuscript reviewer for Oxford University Press, Sage Publications, and academic journals. She has served as a guest lecturer at Hsiuping University, Taichung City Taiwan. Her research interests include qualitative analysis of politics and policy of water resource management, race and ethnicity, rural populations, and post-war American art. Currently, she is curating *Art in*

*Community: The Harold Stevenson Retrospective* at the Museum of the Red River, Idabel OK.

**Dr. Mollie K. Murphy** is an assistant professor of communication studies at Utah State University in Logan, Utah. She holds a Ph.D. in Communication Studies with an emphasis in rhetorical studies from the University of Georgia and both an M.A. and B.A. in Communication Studies from the University of Montana. Her research draws on rhetorical criticism and theory to analyze the communicative strategies of social justice activists. She is particularly interested in issues related to environmental justice, gender, and race. Dr. Murphy is the recipient of several top paper awards from the National Communication Association, and her work has been published in peer-reviewed journals including *Argumentation and Advocacy, Women's Studies in Communication*, and *The Howard Journal of Communications*. At Utah State University, she teaches courses in gender communication, environmental activism, and interpersonal communication.

**Dr. Mehmet Soyer** is assistant professor in sociology at Utah State University. He received his M.A. in sociology at the University of North Texas. Dr. Soyer completed his doctorate in sociology from Texas Woman's University. His major areas of interest are in the subjects of environmental sociology, social stratification, race and ethnicity, and media studies. He has authored a book entitled *Founding Father of Sociology: Ibn-i Khaldun* and has sole authored and co-authored articles in peer-reviewed journals.

**Dr. Sebahattin Ziyanak** is assistant professor in sociology at the University of Texas Permian Basin. Dr. Ziyanak holds a Ph.D. in sociology from the University of North Texas. He received his M.A. in sociology from the University of Houston, Texas and his B.S. in sociology from the Mimar Sinan University in Istanbul, Turkey. He is the recipient of the La Mancha Society Golden Windmill Research Award in 2018. He is the recipient of the Outstanding Excellence in Teaching with the National Society of Leadership and Success in 2018. He is a member of the Advisory Board of the Odessa Links for Odessa Homeless Coalition. He is the President of Peace Academy of West Texas. His fields of research are in the subjects of delinquency, deviance, social organization, social movement, sociology of education, environmental studies, and race and ethnicity. He is also the author of the following books: *Introduction to Sociology* (Cognella: San Diego, CA); (Co-editor with Bilal Sert) *Turkish Immigrants in the Mainstream of American Life: Theories of International Migration*, (Lexington: Lanham, MD); *Analyzing Delinquency among Kurdish Adolescents: A Test of Hirschi's Social Bonding Theory* (Lexington Books: Lanham, MD) and *Crossroad: A Grassroots Organization for the Homeless in Houston* (VDM Publishing: Saarbrücken, Germany).